Cement and Concrete

JOIN US ON THE INTERNET VIA WWW, GOPHER, FTP OR EMAIL:

WWW: http://www.thomson.com
GOPHER: gopher.thomson.com
FTP: ftp.thomson.com
EMAIL: findit@kiosk.thomson.com

A service of I(T)P®

Cement and Concrete

M. S. J. Gani

Faculty of Engineering
Monash University
Clayton, Victoria
Australia

CHAPMAN & HALL
London · Weinheim · New York · Tokyo · Melbourne · Madras

Published by Chapman & Hall, 2–6 Boundary Row, London SE1 8HN, UK

Chapman & Hall, 2–6 Boundary Row, London SE1 8HN, UK

Chapman & Hall GmbH, Pappelallee 3, 69469 Weinheim, Germany

Chapman & Hall USA, 115 Fifth Avenue, New York, NY 10003, USA

Chapman & Hall Japan, ITP-Japan, Kyowa Building, 3F, 2-2-1 Hirakawacho, Chiyoda-ku, Tokyo 102, Japan

Chapman & Hall Australia, 102 Dodds Street, South Melbourne, Victoria 3205, Australia

Chapman & Hall India, R. Seshadri, 32 Second Main Road, CIT East, Madras 600 035, India

First edition 1997

© 1997 M. S. J. Gani

Typeset in 10/12 Times by Florencetype Ltd, Stoodleigh, Devon
Printed in Great Britain by St Edmundsbury Press, Bury St Edmunds, Suffolk

ISBN 0 412 79050 5

Apart from any fair dealing for the purposes of research or private study, or criticism or review, as permitted under the UK Copyright Designs and Patents Act, 1988, this publication may not be reproduced, stored, or transmitted, in any form or by any means, without the prior permission in writing of the publishers, or in the case of reprographic reproduction only in accordance with the terms of the licences issued by the Copyright Licensing Agency in the UK, or in accordance with the terms of licences issued by the appropriate Reproduction Rights Organization outside the UK. Enquiries concerning reproduction outside the terms stated here should be sent to the publishers at the London address printed on this page.

The publisher makes no representation, express or implied, with regard to the accuracy of the information contained in this book and cannot accept any legal responsibility or liability for any errors or omissions that may be made.

A catalogue record for this book is available from the British Library

∞ Printed on permanent acid-free text paper, manufactured in accordance with ANSI/NISO Z39.48–1992 and ANSI/NISO Z39.48–1984 (Permanence of Paper).

Contents

Preface		viii
Acknowledgements		x
1	**History of inorganic (calcareous) cements, mortars and concretes**	1
	1.1 Origins	1
	1.2 Greek and Roman times (~1000 BC to AD~500)	2
	1.3 Middle Ages (AD~400 to AD~1500)	4
	1.4 Eighteenth century	4
	1.5 Nineteenth century	6
	1.6 Twentieth century (to present)	8
	1.7 Cement and concrete in Australia	9
2	**Production of Portland cement**	13
	2.1 Cement kilns	13
	2.2 Raw materials	22
	2.3 Phases in Portland cement clinker	23
3	**Hydration of cement – setting reactions**	36
	3.1 Hydration reactions of individual components in cement clinker	36
	3.2 Morphology of hydrated cement products	41
	3.3 Hydration of cement paste	42
	3.4 Types or Portland cement	46
4	**Mortar**	51
	4.1 Properties of wet mortar	51
	4.2 Strength of mortar	52
	4.3 Mortar mixes	53
	4.4 Sprayed mortar	54
5	**Concrete**	56
	5.1 Effect of the type of cement on strength	57

	5.2	Aggregate and the aggregate/cement bond	57
	5.3	Effects of water/cement ratio and workability on strength	62
6	**Standard tests for cements, cement pastes, mortars and concrete**	70
	6.1	Cements	70
	6.2	Cement pastes	71
	6.3	Mortar	72
	6.4	Concrete	74
	6.5	Non-destructive tests	78
7	**Some additives (admixtures) used in mortar and concrete**	82
	7.1	Accelerators, retardants and stabilizers	83
	7.2	Admixtures for air entrainment	85
	7.3	Additions of pozzolanas	87
	7.4	Superplasticizers	94
8	**High-performance concrete**	100
	8.1	Reduction of the water/cement ratio	101
	8.2	Additions of microsilica	101
	8.3	Production of high-performance concrete	105
	8.4	Structure and properties of high-performance concrete	105
9	**Physical behaviour of concrete after pouring**	109
	9.1	Before setting has commenced	109
	9.2	After setting has commenced	111
10	**Reinforced and prestressed concrete**	118
	10.1	Reinforced concrete	118
	10.2	Prestressed concrete	121
11	**Fibre-reinforced cement and concrete**	128
	11.1	Types of fibre-reinforced cement (mortar)	130
	11.2	Fibre-reinforced concrete	140
	11.3	Some uses of fibre-reinforced cement and concrete	142
12	**Deterioration of cement and concrete**	146
	12.1	Deterioration of concrete	146
	12.2	Corrosion of steel reinforcement	154
13	**Durability and protection of concrete**	159
	13.1	Durability	159
	13.2	Protection	164
14	**Resistance of concrete to fire**	170
	14.1	Fire damage to concrete	170
	14.2	Assessment of fire damaged concrete	174

15	**Special cements and concretes**	**177**
	15.1 High alumina cement	177
	15.2 Fast-setting and hardening cements	183
	15.3 Polymer-modified cements (mortars) and concretes	184
	15.4 Supersulphated cement	189
	15.5 Modification of Portland cement-based materials	190
	15.6 Non-calcareous cements	196
Appendix A: Australian standard Portland and blended cements		**203**
	A.1 General purpose cement	203
	A.2 Special purpose cements	204
Index		205

Preface

This book has been developed from lecture notes prepared for a course on cement and concrete given to materials engineering degree students. The need for such a text was realized when it was found that existing tomes on the subject were written either for civil and construction engineering students, with an emphasis on the design of concrete mixes, quality assurance and testing; or from a chemistry point of view, where the extremely complex chemistry and crystallography of cement was treated in great detail. There was a perceived need for an up-to-date text which concentrated on the morphologies of cement and concrete and their material properties.

The main purpose of this book is to serve as a source of information for an introductory course in cement and concrete materials technology and it contains sufficient material to act as a stand-alone text for such a course. It is also suitable for use as a companion text for civil and construction engineering subjects on concrete.

The primary objective of the book is to give the students an appreciation of the complex nature of a class of materials that is normally taken for granted and a realization that most of the problems frequently reported in the press of failures involving concrete constructions are preventable. To this end an understanding of the material, its production, properties and reactions is essential.

The scope of the book covers the basic science and technology of Portland cement, its manufacture and properties. The production of mortars and concretes from the cement, water and various aggregates are described. Also included are the modification of properties of mortar and concrete by additives and the testing, durability and degradation of concrete. Special emphasis is given to the latest developments in the production of high performance and special cements and concretes.

A systematic approach is taken in presenting the technical information. The book commences with a historical background of the development of cement and concrete, culminating in the development of Portland

PREFACE

cement. This is followed by a description of the production of Portland cement from raw materials. The hydration reactions of the cement are given which lead to the use of the Portland cements in the production of mortars and concretes; test methods used to measure their properties are also presented.

The role of additives to concrete is discussed leading to the production and properties of high-performance concrete. The physical behaviour of concrete from pouring to final set is outlined. Materials and methods used to reinforce, prestress and toughen concrete are also covered. The factors affecting the durability, protection and deterioration of concrete are investigated. Finally, cements and concretes for special applications are described.

In order to make this book useful for the teaching of subjects outside materials engineering, for example in civil and construction engineering, the sections that either depend on a more specialist materials knowledge (e.g. ternary phase diagrams) or are not intrinsically essential for the understanding of the subject (e.g. reference to local Australian conditions) are distinguished by the use of a smaller print. These sections may be omitted from a course without significant loss of overall cohesion and clarity.

Acknowledgements

I am indebted to Julie Fraser for her most valuable help with the diagrams used in the text and to Robert Gani for his many useful suggestions and assistance in proofreading the text.

History of inorganic (calcareous) cements, mortars and concretes

1.1 ORIGINS

The search for materials to bond together stones or bricks in order to fabricate walls, floors and building foundations, together with plasters to cover walls, ceilings, etc., has a long history which extends back to Neolithic times. The use of a lime-based concrete for the construction of polished floors in a Neolithic site in southern Galilee has recently been described by Malinowski and Garfinkel (1991). Carbon dating of grain and other materials found at the site indicate they were made around 7000 BC. The extensive floor area excavated reveals that considerable quantities of lime were use to produce such a large amount of concrete. From this it was inferred that the technology of burning or calcining limestone to form calcium oxide (heating $CaCO_3$ to above 825 °C to form CaO), slaking the lime in water (to form a colloidal suspension of $Ca(OH)_2$), and then mixing the slaked lime with limestone aggregate and water to form concrete, was well known to the Neolithic builders. The lime concrete hardens by the slow reaction of the slaked lime with carbon dioxide to form calcium carbonate. Ceramic remnants of the refractories used to line the kilns used for the calcination of lime have also been identified.

Prior to this excavation, it was thought that the first inorganic cementing material that was used to bond stones together was made from gypsum ($CaSO_4.2H_2O$). It is generally agreed that the Egyptians used gypsum in the construction of the Pyramid of Cheops (2613–2494 BC). Some blocks in the pyramid have a mass of 14.5 tonnes and most of the blocks are bound together with mortars prepared from impure gypsum cement and sand. The cement was prepared by heating the gypsum which causes it to dehydrate to form firstly the hemihydrate ($CaSO_4.½H_2O$) and finally, at higher temperatures, the anhydrous sulphate (Lea, 1970). The dehydrated calcium sulphate was then mixed with sand, and on adding water, rehydration occurred to form interlocking needles of gypsum, which gave the mortar its strength. However, the gypsum deposits were very impure and usually contained calcium carbonate, which means that the use of lime cements cannot be totally discounted, but the higher calcination temperature needed for limestone could be the reason why gypsum was used instead of lime in Egypt where there was a shortage of fuel. Analysis of a mortar that was used for fixing a drain pipe in Egyptian constructions showed it contained 46% gypsum, 41% calcium carbonate and 13% quartz (Neuburger, 1969).

(Modern gypsum cements are still produced by the calcination of gypsum. Plaster of Paris (the hemihydrate) is made by calcining gypsum at ~180 °C; anhydrous calcium sulphate (Keenan's cement) by calcining gypsum at ~550 °C. Both of these materials harden by the hydration of the compounds to form the dihydrate, but the rates of hardening and the morphologies of the hydrated crystalline products are different. Plaster of Paris sets rapidly to form a porous matrix; Keenan's cement hardens at a much slower rate, and the interlocking nature of the hydrated crystals result in the formation of a less porous, stronger cement.)

According to Morris (1991) the debate about the materials used to construct the pyramids has been widened by Davidovits who claims that the blocks used to construct the pyramids were not made of quarried limestone, but were actually made from concrete that was cast in place. Davidovits suggests that a limestone concrete was used which was made from a mixture of limestone, lime, kaolinite (an aluminosilicate clay mineral), natron (sodium carbonate, $Na_2CO_3.10H_2O$) and water. This claim has been questioned by Campbell and Folk (1991).

In the Middle East some additional evidence has been found of the early use of lime mortars. Some of the bricks found in the ruins of a palace are said to be laid in lime mortar. These bricks were stamped with the name of Nebuchadnezzar (~630 to 562 BC). In the nearby city of Ur, part of a temple had bricks laid in a mortar made from lime and ashes (Draffin, 1976).

1.2 GREEK AND ROMAN TIMES (~1000 BC TO AD~500)

Lime cement based mortars were used by the Greeks. Sand was normally added to the lime cement to produce the mortar. Lime mortar is sometimes referred to as 'air-mortar' since it hardens after drying by reaction with carbon dioxide in air (Draffin, 1976).

The Romans perfected the art of the preparation of lime mortars. Examples of Roman brickwork held together with lime mortars still exist.

GREEK AND ROMAN TIMES (~1000 BC to AD~500)

The excellence and durability of the Roman mortars were due to the care taken in the mixing and ramming of the mortar, and not in the composition of the lime or any secret used in the slaking process. The major difference in the lime mortars used by the Romans and those prepared by the Greeks was in the amount of sand added to the mixture. The Greeks used a 1:7 sand to lime ratio whereas the Romans used a ratio of 2:1 or even 3:1 in their mixture (Neuburger, 1969).

Hydraulic setting cements were first developed by the Greeks and Romans. These cements exhibit an increase in strength when stored under water after setting. They were made by the addition of volcanic ash (*pozzolana*) to the slaked lime and sand to produce a mortar which possessed superior strength when compared with the lime/sand mixture and when set was resistant to the action of water. (The original volcanic ash used as a pozzolanic material was produced by two violent volcanic eruptions in the Mediterranean, the first being in the Aegean island of Thera, now known as Santorin in Greece, and the second in AD 79 when Mt Vesuvius erupted on the bay of Naples in Italy (ACI Committee 232 Report, 1994).) The *pozzolana*-lime mixture was discovered by the Greeks sometime between 700 and 600 BC, this was later passed on to the Romans who used it in mortar and concrete in about 150 BC (Kirby *et al.*, 1956, p. 63). Pozzolanic cement mortar which has lasted over the centuries was used in the construction of the Colosseum in Rome (Drysdale *et al.*, 1994). The hydraulic cement produced from this material is known as 'Roman' cement.

Marcus Vitruvius Pollio (Vitruvius Pollio) recorded in his treatise *De Architectura* written in about 27 BC, 'There is a species of sand which, naturally, possesses extraordinary qualities. It is found under Baiae and the territory in the neighbourhood of Mount Vesuvius; if mixed with lime and rubble, it hardens as well under water as in ordinary buildings' (Lea, 1970).

Other pozzolanic materials were found in Sicily and the Campagna. In Germany and the low countries the material was present in ancient volcanic deposits as a consolidated rock which when crushed produced trass (Kirby *et al.*, 1956, p. 63). The Romans also found that finely ground tiles or pottery added to the mortar had a similar effect. According to Vitruvius, 'If to river or sea sand, potsherds ground and passed through a sieve, in the proportion of one-third part, be added, the mortar will be the better for its use' (Lea, 1970).

Hydraulic setting concrete, which is a mixture of aggregate (gravel), sand and Roman cement, has been used since Roman times. Much of the best concrete used broken brick aggregate, lime and pozzolana. In large works, porous volcanic tuff was used to replace the bricks as the aggregate in the concrete mix in order to decrease the weight of the structure. The Pantheon in Rome, a circular temple which was built during the reign of the Emperor Hadrian in about AD 120 and is still standing, is constructed of such concrete (Kohlhaas, 1983). The walls are 6.1 m thick and are made of volcanic tuff and pozzolana concrete thinly faced with brick. The dome's 43.4 m span is cast solid in lightweight concrete containing pumice and pozzolana (Lea, 1970; Straub, 1960). The Roman concrete (*opus caementicium*) differed from modern concrete in that the aggregate (*caementa*) and the mortar were not pre-mixed but laid separately in horizontal courses (Harries, 1995). This was necessary since the size of the aggregate (50–150 mm) was too large to be mixed manually. Harries (1995) also describes how the nature of the aggregate was graded according to the position in the building, with dense, strong basalt being used in the footings, clay brick in the walls to light weight pumice in the dome. The Basilica of Constantine in Rome which had a vaulted nave 25.3 m wide was also cast in concrete. Concrete was also used by the Romans in the construction of the aqueducts at Segovia in Spain and Pont du Garde in France. In the tomb of the Empress Helena, located near Rome, the weight of the concrete used in its construction was reduced by embedding hollow pots into the concrete. The roof of the Baths of Caracalla in Rome had metal bars incorporated in their structure (Taylor, 1969).

In an interesting account of concrete constructions in early Rome, Harries (1995) makes the point that the use of concrete in construction had the advantage over the use of stone in that, under supervision, large numbers of unskilled workers (slaves) could be used instead of the highly skilled craftsmen and stone masons.

1.3 MIDDLE AGES (AD~400 TO AD~1500)

After the decline of the Roman Empire, there was a gradual decline in the quality of the cement, mortar and concrete used in building construction. It was not until the twelfth century that the quality started to improve, when care was again taken in the calcination of the lime and the mixing of mortars (Lea, 1970). Concrete was used as a construction material throughout the Middle Ages, mainly for churches and castles, but, like the mortar, the concrete was generally of inferior quality when compared with that made by the Romans.

1.4 EIGHTEENTH CENTURY

In more recent times, the most important advance in the knowledge of cement was made by John Smeaton who was given the task of erecting a new lighthouse on the Eddystone Rock. This rock, which is covered by sea at high tide, is situated in the main shipping lanes off Plymouth in the English Channel. The previous structures had both failed, one built by Henry Winstanley in 1699 had been destroyed by a storm in 1703, and the replacement built by John Rudyerd in 1709 was destroyed by fire in 1755. Smeaton investigated the cementing properties of various mortars, made from lime obtained from various locations, and discovered that the best mortars were made from the calcination of limes that contained considerable proportions of clay minerals (argillaceous lime). This was the first occasion that the importance of clay mixed with the lime had been recognized in the formation of a hydraulic setting cement. It was found that limes that did not dissolve completely in nitric acid (clay being insoluble in the acid) possessed good hydraulic properties (Kohlhaas, 1983). The cementitious agent Smeaton finally used was made from such a clay containing lime which was mixed with an equal quantity of pozzolana (Lea, 1970). The lighthouse that he constructed stood for 123 years until 1879 and only failed when its foundations were undermined by the sea. Smeaton's conclusions about the importance of the presence of clay were not published until after his death in 1792. Smeaton was the first to call himself a civil engineer (as distinct from a military engineer). In the preface to his book, *Hydraulischen Mörtel*, W Michaëlis stated in 1869 (in translation):

> A century has elapsed since the famous Smeaton completed the building of the Eddystone Lighthouse. Not only the seafaring but for all humanity stands as a true signal of blessed work, a light in the dark night. From the scientific point of view it illuminated the darkness of nearly 2000 years.
>
> The errors which came to us from the Romans and which were shared even by the excellent Belidor, were dispersed.
>
> The Eddystone Lighthouse is the foundation upon which our knowledge of hydraulic mortars has been built and is the chief pillar of modern construction. Smeaton freed us from the shackles of tradition by showing us that the purest and hardest limestone is not the best, at least for hydraulic purposes, and that the source of the hydraulicity of lime mortar must be sought in the argillaceous admixtures (Draffin, 1976).

This cement was called 'Roman cement' although it in no way resembled the true Roman cement, except for its hydraulic setting reactions. A patent on the hydraulic setting cement made from argillaceous lime (septaria) was taken out in England by James Parker in 1796. The calcining temperature was higher than that used for calcining lime, but not high enough to vitrify (partially melt) the argillaceous lime nodules. The fired product was ground to a powder before use. Any material that had accidentally fused was picked out and discarded (Draffin, 1976). This was probably because any partially fused clinker was found to be extremely hard and therefore very difficult to grind into a fine powdered cement (Klemm, 1989).

1.5 NINETEENTH CENTURY

In 1818, Vicat, in France, prepared an artificial 'Roman cement' by calcining an artificial mixture of limestone and clay (as distinct from using naturally occurring clay/lime mineral admixtures). This material was the forerunner of Portland cement. He also invented the Vicat needle which is still used today to determine the setting rate of the cement. Vicat's investigations into the effect of the composition of the raw materials used in making the cement on the properties, revealed that limestones which form hydraulic limes contained silica, alumina, magnesia, manganese and iron to the extent of from one-fifth to one-quarter of the total weight. He found that no perfectly hydraulic mortar could exist without silica, and that hydraulic limes contained clay (aluminosilicate minerals). This confirmed Smeaton's observations that a lime or cement with hydraulic properties contains lime, silica and alumina (Draffin, 1976). Both Vicat and Joseph Aspdin (a builder from Leeds, England) were manufacturing artificial hydraulic cement on a commercial scale at the beginning of the nineteenth century by reacting carefully proportioned mixtures of clay and limestone at elevated temperatures (Young, 1995).

Portland cement was patented in 1824 by Joseph Aspdin. It is said that before the process was patented, Aspdin used to put brightly coloured, but useless salts, into his kiln in order to deceive his rivals (Young, 1995). The cement was of rather poor quality due to the low calcination temperature used. The product was called Portland cement because the set product bore some resemblance to Portland stone. Unlike the Roman mortars, which would resist water only after setting had commenced, Portland cement would set under water. Marc Isambard Brunel (1769–1849), father of Isambard Kingdom Brunel, employed Portland cement for the mortar used in the construction of a brick lined tunnel under the Thames river in London (Young, 1995). The work started in 1825 and was finished in 1843 when Brunel was 74 years old. Brunel's use of the Portland cement developed by Aspdin aided the acceptance of the material (Upton, 1975). The tunnel was originally used for road traffic, then for a railway and is now part of the London underground system (Kirby et al., 1956, p. 487). It was Aspdin's son, William, who improved the quality of the cement by using higher calcination temperatures at which the cement clinker partially melted (Kohlhaas, 1983). This superior cement was used in the construction of the Houses of Parliament in London in 1840 to 1852. The first extensive use of Portland cement was in the construction of the London sewerage system in 1859 to 1867. This resulted in an increase in popularity of the product and quantities of the cement were exported to other countries. In 1896 nearly 3 million barrels were exported to the USA (at this time the local production of 'natural' (Roman) cement in the USA was 0.8 million barrels/year) (Draffin, 1976).

According to Kirby *et al.* (1956, p. 476), the French were the chief pioneers in concrete constructions in the nineteenth century. Poirel, in 1833, used massive precast blocks (16 000 blocks weighing about 20 tonnes each) in the construction of jetties at Port Said which is situated at the entrance of the Suez Canal. Three miles of concrete bridges were used in the construction of the Vanne Aqueduct through the Fontainbleau Forest to bring water to Paris in 1870.

Despite the development of Portland cement, and its use in 1838, natural cement made from limestones containing clay minerals continued to be used until the end of the nineteenth century. Cement works were established near deposits of natural cement rock (argillaceous limestones and gravels) both in Europe and America. This natural cement was used in America in the construction of the Erie Canal (Upton, 1975). In America, the production of natural cement reached its peak in the 1890s, only to be overtaken by Portland cement production to such an extent that the amount of natural cement produced 20 years later was negligible compared with that of Portland cement (Draffin, 1976).

The modern, high-temperature process, of making Portland Cement was discovered in 1844 by Isaac Johnson who heated the ingredients to a temperature at which they partially melted. This not only shortened the calcining time, but also resulted in the production of a cement with more reliable setting and strength properties. The processing of the hard clinker produced by this process to a fine powdered cement was simplified by the use of a jaw crusher (Blake's Stonebreaker) which was introduced into England in 1862 (Klemm, 1989).

It was recognized that the amounts of clay and lime had to be carefully proportioned in order to produce Portland cement; however the crude volumetric ratios used early in the nineteenth century left much to be desired. It was only late in the century (1880) that chemical analysis of the raw materials was being carried out on a routine basis. It was in 1887 that Le Chatelier published his doctoral thesis on the chemistry of cement and gave the upper and lower limits of the amount of lime that should be used to produce a good cement (Klemm, 1989).

The production of modern Portland cement resulted in the widespread use of concrete containing this cement by 1880. Steam-driven concrete mixers were also in use by this time (Young, 1995). The use was further increased by the introduction of reinforced concrete in which the high compressive strength of concrete is combined with the high tensile strength of steel. Reinforcement of concrete was patented by an Englishman, Wilkinson, in 1855. From these beginnings, a large number of patented reinforcing bars and systems were rapidly developed and used in concrete constructions (Low, 1991). The first building in Britain with a reinforced concrete frame was built by Hennebique in 1897 at Weaver's Mill, Swansea. He also built a reinforced concrete bridge in Scotland (Mays, 1992).

Steel mesh reinforced cement mortar (ferrocement) was first used in France by Lambot, who exhibited a boat made from ferrocement at the Paris International Exposition in 1855 (Low, 1991). One of the original Lambot boats which was found when a lake was drained is described by Fisher Cassie (1967). The boat is 3.6 m long, 1.35 m across the beam and the sides are 30–40 mm thick. The 'Monier trellis' was patented in 1865 by Monier (a Parisian gardener). The trellis consisted of a set of reinforcing bars which was used in the construction of flower tubs from cement mortar.

In the nineteenth century there was an increasing amount of iron produced in blast furnaces, and, in order to remove some of the impurities out of the pig iron, it was necessary to add limestone to the raw materials (iron oxide and coke). During the reduction of the iron oxide by coke at high temperatures, the limestone decomposed to calcium oxide, which then reacted with silica, alumina and other oxide impurities to form a molten slag which floated on top of the molten iron in the furnace. Approximately equal volumes of iron and slag were produced, and this meant that with the increase in iron production, large amounts of slag were also made as a waste product. If this slag was rapidly cooled, it formed a glass, which, when finely ground to a powder, had hydraulic properties, i.e. it acted as a slow setting cement when mixed with water. The rate of setting was increased in the presence of calcium hydroxide. The cementitious potential of blast furnace slag was discovered by Emil Langen in Germany in 1862 (Higgins, 1991). It was found that when the blast furnace slag was mixed with Portland cement, sufficient calcium hydroxide was produced by the setting Portland cement to activate the blast furnace slag. The slag, being a waste product, was much cheaper than Portland cement, and could be used to substitute for some of the cement in a concrete mix, with the result of a lowering of cost. There were also improvements in other properties of the concrete which resulted from the slag additions. The first commercially available mix was produced in Germany in 1865 (Higgins, 1989). In 1880 the potential for the use of Portland cement mixed with blast furnace slag in aggressive soil conditions was realized (Higgins, 1991). The blast furnace slag admixture was used in 1889 in the construction of the Paris Underground Metro system.

1.6 TWENTIETH CENTURY (TO PRESENT)

Early in the twentieth century concrete made with Portland cement which was used for the construction of railway tunnels in France failed. The failure was found to be due to attack on the cement by ground-water which contained dissolved sulphates. To counteract this, a sulphate resistant high alumina cement (HAC) was produced by melting bauxite (aluminium hydroxide) with lime. The product was patented in 1908 and was a hydraulic setting cement which was not only resistant to sulphate attack, but also set and hardened very rapidly. Concrete made from high alumina cement was widely used in the building industry due to its early strength development, which meant that formwork, which was used to support the concrete while the cement set, could be removed quickly. But later HAC concrete was found to fail if exposed to a combination of high humidity and temperatures, and its use is now limited (Neville, 1975).

Reinforced concrete was used to construct the Risorgimento Bridge over the Tiber river at Rome in a single span (Kirby *et al.*, 1956, p. 477). The strength and stiffness of the bridge was tested by marching 110 soldiers over the bridge in double-quick time and then driving three 15 tonne rollers abreast forwards and backwards across the bridge. The largest deflection measured was 2.5 mm.

Most of the concrete used in construction for over a century was made from a mixture of cement, aggregate and water. The water played a large role in determining the strength of the concrete. There had to be sufficient water to ensure that hydration of the cement occurred and also to impart to the fresh concrete sufficient workability for its placement. This normally meant that excess water had to be used, and this excess water controlled, to a large extent, the strength and durability of the concrete. Since the 1980s there have been two major developments which have led to the production of high-strength concrete by the use of various admixtures. The first was the improvement of the mix design by the extension of the grain size range used in the concrete mix by the addition of extremely fine nano-sized particles (e.g. microsilica) which filled the voids that had existed between the larger cement particles. This resulted in a more compact mixture, and increased the strength. The second was the addition of organic compounds which increased the fluidity of the fresh concrete mix, which meant that much less water could be used in the mix. These additives have resulted in the compressive strength of concrete used in the construction industry increasing from 20–40 MPa to 80–120 MPa. This, in turn, has led to major changes in the design and methods of construction of buildings. Research is now under way into optimizing a combination of the two improvements in order to obtain a further increase in strength.

1.7 CEMENT AND CONCRETE IN AUSTRALIA

Cement manufacture in Australia was first attempted in 1859 in Victoria, but competition from imported cement resulted in the commercial failure of the early projects (Lewis, 1988). It was not until 1892 that the first sustainable cement production plant was built in South Australia. Plants were then set up in New South Wales, Tasmania and Victoria. There were problems with the production of cement of consistent quality, and this became of major importance with the introduction of reinforced concrete soon after 1900. Rotary kilns for the production of cement clinker were introduced in the early 1900s. With the outbreak of the First World War in 1914 the importation of cement from overseas ceased and locally produced cement had to be used. Australian standards for cement were introduced in 1925 and adopted in 1926.

Reinforced concrete construction commenced in Australia in the early 1900s. Sir John Monash was involved in the design and/or construction of many of the first reinforced concrete constructions in Melbourne including the Public Library, the Bank Place Chambers and the Anderson Street bridge. The dome of the Public Library, with a span of 34.8 metres, was the largest reinforced concrete frame and dome in the world at that time in 1909 (Lewis, 1988). Monash also introduced the use of reinforced concrete into South Australia, and established his own company (Monier) in Victoria.

Concrete that used to be considered as 'high-strength' (50–60 MPa compressive strength) has been available in Melbourne since the 1970s (Burnett, 1989), but had to be placed by crane rather than by pump (e.g. the Rialto building and the Collins Place project). The use of chemical admixtures has resulted in a pumpable mix being used in the Melbourne Central, Bourke Place and 550 Collins Street developments. This resulted in considerable savings in building costs. Similarly, silica fume (microsilica) has been used in concrete for recent constructions in Melbourne (Burnett, 1991). A 90 MPa pumpable mix was used for some of the columns supporting the upper deck of the new grandstand at the Caulfield Race Course. This meant that the requirement that the columns be as slim as possible, so as not to obstruct the view of the finishing straight, could be met. A comprehensive listing of projects in which high-performance concrete has been used in Australia in recent years is given by Papworth and Burnett (1993).

QUESTIONS

1. What is the difference between cement, mortar and concrete?
2. Write down the equation for the carbonation reaction of a lime mortar. Would you expect the rate of carbonation of a lime mortar set between bricks to be linear? List the factors that you consider could be important in the carbonation process.
3. How are gypsum mortars produced and why do they set at a much faster rate than lime mortars?

4. Given that the Egyptians will not permit the removal of any material that was used in the construction of the pyramids, or any destructive testing in situ, describe how would you would attempt to determine whether the blocks from which the pyramids were built were cast using a limestone concrete as claimed by Davidovits.
5. What was the major difference between the cement and concrete used by the Romans and those used by their predecessors?
6. What is the difference between the 'Roman Cement' patented by Parker in 1796, and the cements that were developed in Roman times?
7. What are the essential ingredients used to make Portland cement and what was the unique feature of mortar and concrete made from this cement?
8. Why was high alumina cement developed, what are its desirable properties and what is the reason for it not being widely used as a construction material today?

References

ACI Committee 232 Report (1994) Use of natural pozzolans in concrete, *ACI Materials Journal*, **91** (4), 410–26.

Burnett, I. (1989) High-strength concrete in Melbourne, Australia. *Concrete International*, **11** (4), 17–25.

Burnett, I. (1991) Silica fume concrete in Melbourne, Australia, *Concrete International*, **13** (8), 18–24.

Campbell, D. H. and Folk, R. L. (1991) The ancient Egyptian pyramids – concrete or rock? *Concrete International*, **13** (8), 28, 30–9.

Draffin, J. O. (1976) A brief history of lime, cement, concrete and reinforced concrete. In H. Newlon Jr. (ed.), *A Selection of Historic American Papers on Concrete, 1876-1926*, American Concrete Institute, Detroit, pp. 3–38.

Drysdale, R. G., Hamid, A. A. and Baker, L. R. (1994) *Masonry Structures: Behaviour and Design*, Prentice-Hall, Englewood Cliffs, New Jersey, p. 6.

Fisher Cassie, W. (1967) Lambots' boats. *Concrete*, **1** (11), 380–2.

Harries, K. A. (1995) Concrete construction in early Rome. *Concrete International*, **17** (1), 58–62.

Higgins, D. (1989) Development and trends in the use of GGBS. *Concrete*, **23** (7), 35–7.

Higgins, D. (1991) The historical development of GGBS. *Concrete*, **25**, 7–9.

Kirby, R. S., Withington, S., Darling, A. B. and Kilgour, F. G. (1956) *Engineering in History*, McGraw Hill, New York.

Klemm, W. A. (1989) Cementitious materials: historical notes. In J. P. Skalny (ed.), *Materials Science of Concrete I*, The American Ceramic Society, Westerville, pp. 1–26.

Kohlhaas, B. (1983) *Cement Engineers' Handbook*, 4th English edn, Bauverlag, Weisbaden, p. 103.

Lea, F. M. (1970) *The Chemistry of Cement & Concrete*, 3rd edn, Edward Arnold, London, Ch 1.

Lewis, M. (1988) *200 Years of Concrete in Australia,* The Concrete Institute of Australia, North Sydney.

Low, R. E. (1991) Reinforced concrete at the turn of the century. *Concrete International,* **13** (12), 67–73.

Malinowski, R. and Garfinkel, Y. (1991) Prehistory of concrete. *Concrete International,* **13** (3), 62–8.

Mays, G. C. (1992) The behaviour of concrete. In G. C. Mays (ed.), *Durability of Concrete Structures,* E. & F. N. Spon, London, New York, pp. 3–9.

Morris, M. (1991) The cast-in-place theory of pyramid construction. *Concrete International,* **13** (8), 29, 39–44.

Neuburger, A. (1969)*The Technical Arts and Sciences of the Ancients,* (translated by H. L. Brose), Barnes & Noble, New York, pp. 403–8.

Neville, A. (1975) *High Alumina Cement Concrete,* John Wiley & Sons, New York.

Papworth, F. and Burnett, I. (1993) High performance silica fume concrete application, in *Concrete 63, 16th Biennial Conference of the Concrete Institute of Australia,* pp. 274–301.

Straub, H. (1960) *A History of Civil Engineering* (translated by E. Rockwell), Leonard Hill, London, Ch 1.

Taylor, W. H. (1969) *Concrete Technology and Construction,* 3rd edn, Angus & Robertson, Sydney, p. 1.

Upton, N. (1975) *An Illustrated History of Civil Engineering,* Heinemann, London.

Young, J. F. (1995) Engineering advanced cement-based materials for new applications. In A. Aguado, R. Gettu, and S. P. Shah (eds), *Concrete Technology: New Trends, Industrial Applications,* E. & F. N. Spon, London, pp. 103–12.

Production of Portland cement   2

The production of Portland cement involves the firing (burning) of calcareous material (normally limestone, sea shells etc.) with an argillaceous material (clay, an aluminosilicate). The solid raw materials are crushed and mixed in ball mills, and then heated in a kiln to about 1500 °C. The firing results in the formation of a clinker which consists of a number of compounds which set or harden when the clinker is ground to a fine powder (cement) and then mixed with water.

2.1 CEMENT KILNS

In Roman times, vertical furnaces were used to burn lime and such kilns were still being used for lime mortar production well into the nineteenth century. Bottle and shaft kilns were the first types employed in the production of Portland cement. These kilns were manually charged and controlled and the irregular operation often resulted in the production of cement clinker with unpredictable and inferior properties. The capacity of the kilns were also limited to less than 300 tonnes per day (Kohlhaas, 1983, p. 307).

The development of the rotary kiln, which is now normally used in the production of Portland cement, probably started about 1877 in England, but was not patented until 1885 by Frederick Ransome (Peray, 1986, p. 3). The rotary kiln is essentially a large refractory lined steel tube inclined at about 3° to 5° to the horizontal. At the lower end of the tube is a burner. The raw materials are fed into the other end. The kiln is rotated slowly (typically 20–86 rph) and the small incline results in the materials passing down the kiln towards the burners. Chemical and physical reactions take place during the passage of the raw materials through the kiln, and clinker (the reaction product) emerges at the burner end. It

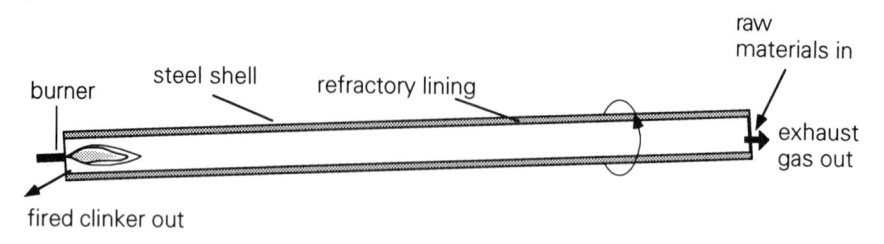

Figure 2.1 Schematic diagram of a rotary cement kiln

is essential that there be an oxidizing atmosphere in the kiln in order to produce cement clinker. Clinker produced under such conditions is greyish-green in colour, whereas if reducing conditions are used the clinker is brownish and produces cement with lower strength and faster setting properties (Kohlhaas, 1983, p. 119). The early Ransome kilns were about 0.45 m diameter and 4.6 m long. The size of the dry process kilns used today is typically 4 m diameter and 70 m long. Modern rotary kilns can produce between 6000 to 8000 tonnes of cement clinker per day (Kohlhaas, 1983, p. 307).

A diagram of a rotary kiln is shown in Figure 2.1.

There are three major processes used for the production of Portland cement in the rotary kiln, these are the wet, semi-dry and dry processes.

2.1.1 Wet process

In the wet process the raw materials are ground and mixed with water and the resultant slurry is fed into the kiln. The average moisture content of the slurry is about 37–39% and this moisture has to be driven off prior to the calcining of the cement (Bulavin *et al.*, 1986, p. 215).

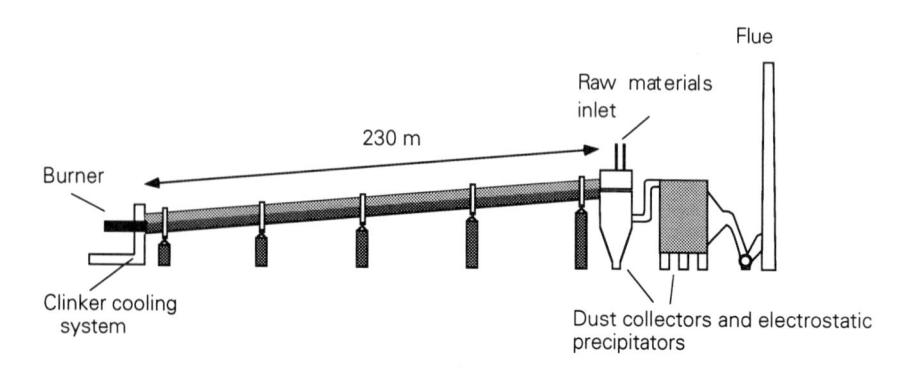

Figure 2.2 Schematic diagram of a wet process kiln

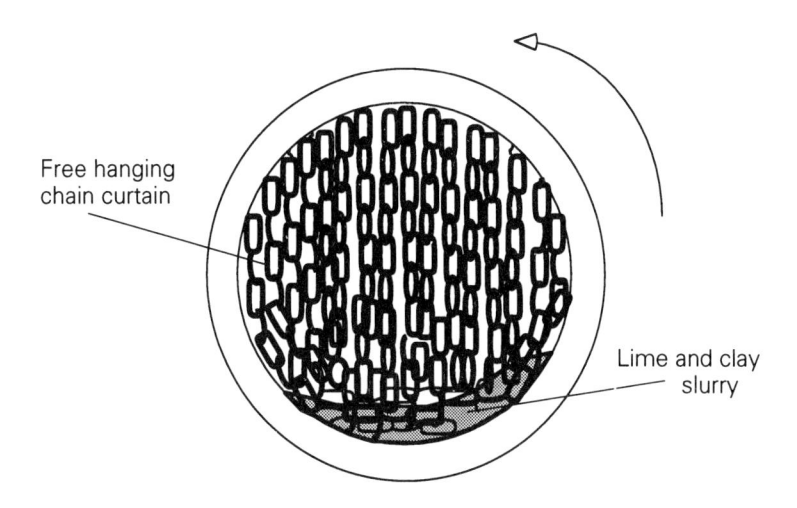

Figure 2.3 Chain curtain for heat transfer from the kiln gases to the feedstock

A diagram of a wet process kiln is shown in Figure 2.2.

To improve the heat transfer between the hot gases of the kiln and the incoming slurry materials, metallic chains can be used as a heat transfer medium. The chains absorb heat from the hot gas from the burner which passes above the slurry and transmits the heat to the kiln feed as the kiln rotates and the hot chains mix with the kiln feed. In the wet process, the chains are used to drive off the free water from the slurry. The slurry enters the chain section at 38 °C and leaves at about 90–100 °C with a moisture content of 6–12%. There are problems with the wet slurry adhering to the chains. As the moisture evaporates, the flow properties of the slurry changes to such an extent that the slurry is converted into granules. Such a chain curtain is shown schematically in Figure 2.3.

There are many other kinds of heat exchangers of various designs and materials (both metallic and ceramic). Cycloidal metallic heat exchangers were developed in the USSR which minimized the problems of clogging that occurs with other designs. They are said to increase the productivity of the kiln by 5% and reduce the unit fuel consumption by up to 7% (Bulavin et al., 1986, p. 230).

Temperature profiles of the gas and solid materials in the wet process rotary kiln are shown in Figure 2.4.

In the clinkering zone, where the reactants partially melt, a dam is used to ensure that the reactants stay in the hot zone long enough for the desired chemical reactions to take place. The clinker is then cooled in planetary coolers in which the heat given off by the hot clinker is used to preheat air which is then used in the burners and precalciner.

PRODUCTION OF PORTLAND CEMENT

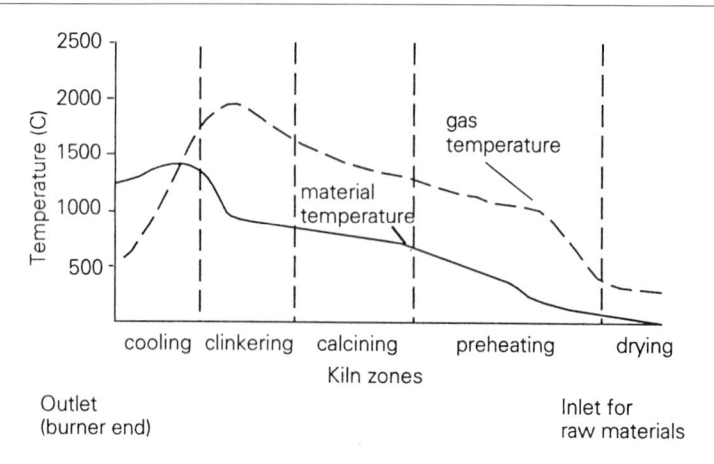

Figure 2.4 Temperature profiles of the gas and solid materials

2.1.2 Semi-dry process (grate process kilns or Lepol kilns)

In the semi-dry process, the feed material contains 10–15% moisture and is pelletized into small nodules. These nodules are dried and partially calcined by the hot exit gases from the kiln before the pellets enter the kiln. The drying can be done using a grate, as shown in Figure 2.5. One advantage of the semi-dry grate process kilns is the uniform size of the clinker that is produced in the kiln from the nodules. This makes the subsequent grinding of the clinker into cement powder easier. Cyclones can also be used to dry the nodules with the hot gases that emerge from the kiln. Chain curtains are also used to transfer heat to the incoming partially calcined feed. Because the feed is dry when it enters the chain section clogging is not a major problem.

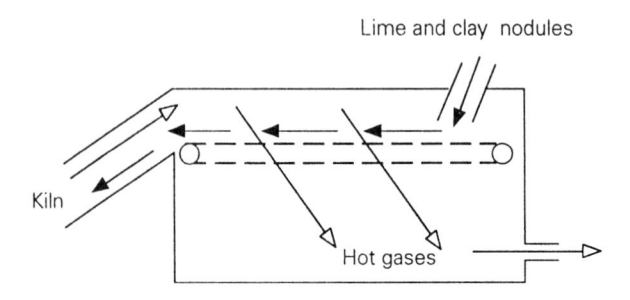

Figure 2.5 Lepol grate for drying and partially calcining nodules prior to entry into the kiln

CEMENT KILNS

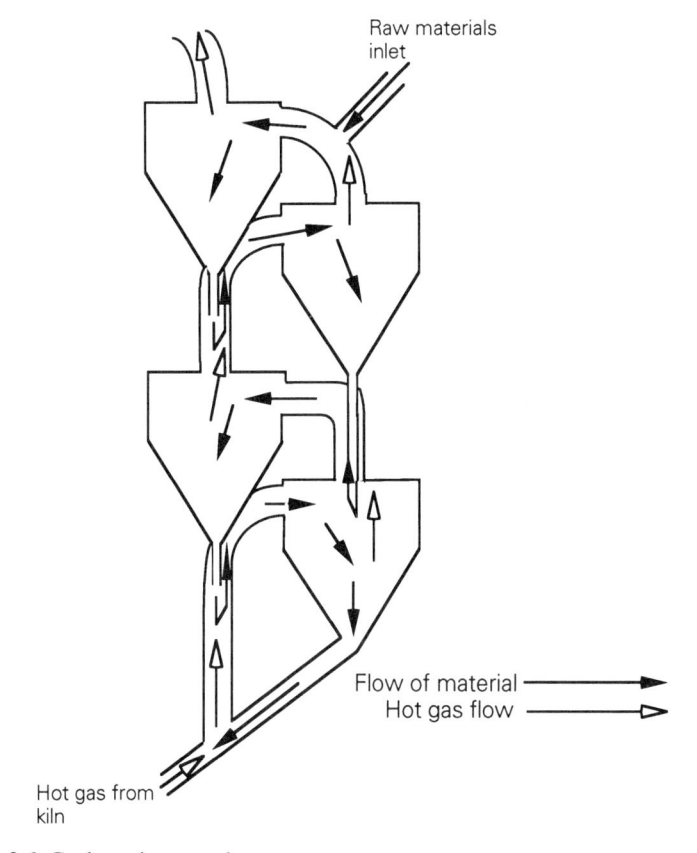

Figure 2.6 Cyclone heat exchangers

2.1.3 Dry process

The dry process is one in which raw materials are fed into the kiln in dry powder form. Cyclones are used to preheat and partially calcine the dry raw materials by the use of the hot kiln gases before the materials enter the rotary kiln, as shown in Figure 2.6. The use of cyclones to preheat the raw materials commenced in the 1950s. In the chain heat exchanger section, the feed enters at 50 °C and leaves at 730 °C, the gases enter the section at 815 °C and leave at 450 °C (Peray, 1986, pp. 8–9).

The production rate of the cement kiln can be further increased if the incoming feed is not only dried and preheated, but it is also precalcined before it enters the kiln. This was done in the 1970s by adding a burner and furnace between the preheating section and the kiln, as shown in Figure 2.7. This allows the kilns to be of smaller diameter without sacrificing kiln output. A kiln with a precalciner will produce 50–70% more

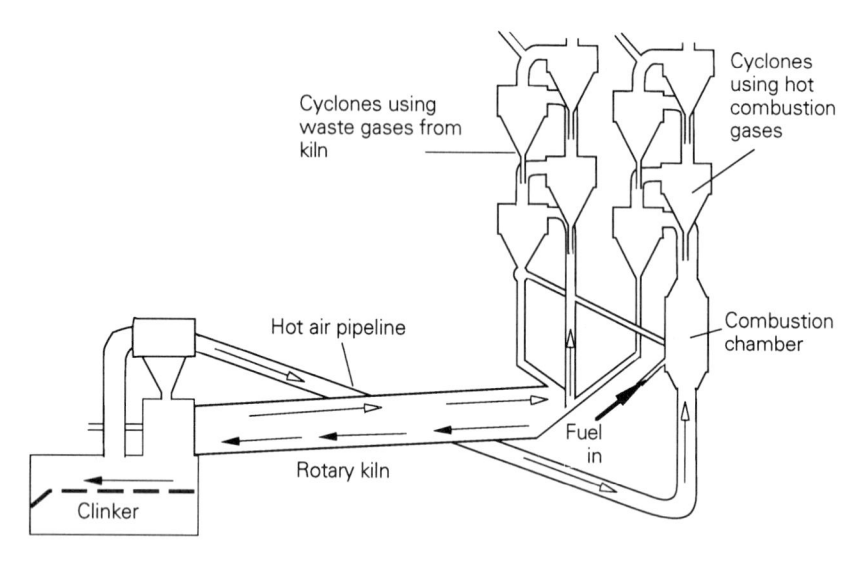

Figure 2.7 Cyclone heat exchangers and precalciner (redrawn from Bulavin *et al.*, 1986, p. 237)

clinker than a conventional preheater kiln of equal diameter. It also serves to reduce the heat load in the hot zone of the kiln and results in extended refractory life in the kiln.

The kiln has an outer steel shell 100 mm thick which rests on six cast steel rollers, each weighing about 100 tonnes. The relative size and production capacity of wet and dry process kilns is given in Table 2.1.

A schematic comparison of the relative size of the kilns for the various processes is shown in Figure 2.8.

2.1.4 Fuel

The kiln is fired using either powdered coal, oil, natural gas or, more recently, waste materials.

In the case of coal, the coal ash that is produced becomes incorporated into the cement. The amount and composition of the coal ash must be

Table 2.1 Comparison of wet and dry process kilns

	Wet	Dry
Kiln diameter (m)	5.0	4.0
Kiln Length (m)	165	70
Clinker, tonne/day	1050	2300

CEMENT KILNS

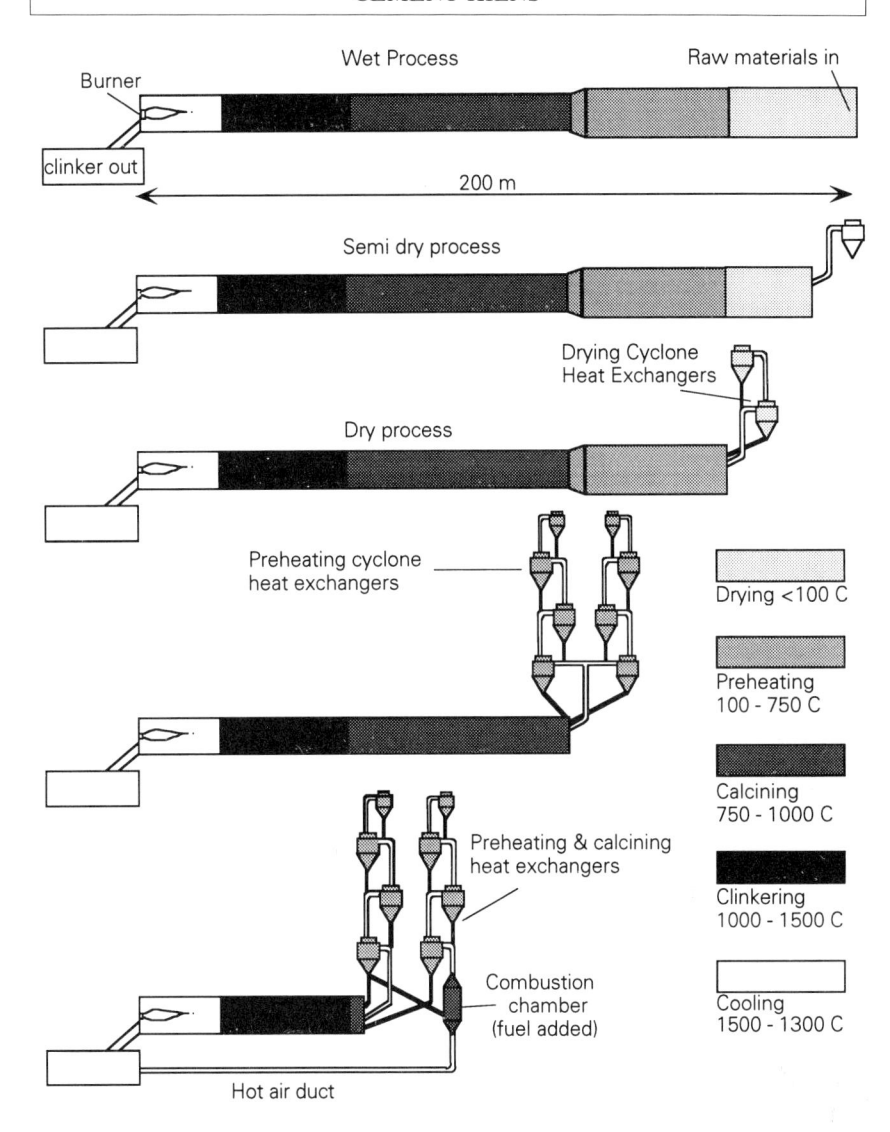

Figure 2.8 Relative size of rotary cement kilns (redrawn from Johansen, 1989)

taken into account when calculating the composition of the feed material. The quantity of ash produced by burning coal varies widely: black coals contain 5–20% ash, and brown coals often contain more than 30% ash. The composition variations are also large, so it is desirable that coal from only one source be used once the coal has been characterized.

The use of natural gas as a fuel has the advantage over coal as it needs no preparation before use (drying, grinding, or preheating). The combustion of the gas is much cleaner and there is little or no need for primary air so the hot secondary air can be used for combustion in the kiln (Peray, 1986, p. 41).

In recent years there has been a trend towards cement kilns using combustible waste materials as fuels. These are materials such as used oil and tyres, municipal solid wastes and waste wood. Cement kilns have been proven to be very efficient in waste disposal, the high gas temperatures (2000 °C), long residence time at high temperature, large thermal mass, alkaline environment and the incorporation of any ash residue in the clinker all contribute to the process. The longer residence times and higher temperatures provided by the cement kilns, when compared to conventional incinerators used for the destruction of waste materials, lead to assured waste destruction (Morton, 1991). The organic part of the waste is burnt in the oxygen rich flame of the kiln and the inorganic part is incorporated in the cement clinker. Wastes with low calorific value can be used in the precalciners, where the high ash content of these fuels can be well mixed with the incoming raw materials in the cyclones (Osbaeck, 1994).

Cement kilns are also being used for the disposal of hazardous wastes such as paint residues, chemical process sludges, oil refinery sludges, chlorinated hydrocarbons etc. With such wastes it is necessary to install special equipment for pretreating the waste, for collection of dust residues and for monitoring potentially corrosive chemicals such as chlorine (Morton, 1991).

The kilns are now in competition with incinerator interests for high calorific value wastes, and it is predicted that this competition might result in the closure of some kilns that rely heavily on these wastes to remain economical (Redeker, 1994). The safety of the cement kilns is also being questioned, despite the results of environmental studies carried out in Midlothian, Texas, which is said to have the highest concentration of cement plants for burning hazardous wastes in the world. The studies concluded that there were no environmental problems with kiln emissions (Kelly *et al.*, 1993).

2.1.5 Refractories

In order to protect the outer steel shell of the kiln from high temperatures within the kiln the whole of the kiln is lined with refractory materials which insulate the shell from both the high temperatures and the potentially corrosive reactions that take place within the kiln. These refractory materials are normally in the form of brick linings, but castable refractory cements and concretes can also be used.

In the cooler zones, aluminosilicate based refractories are used (25–80% Al_2O_3), but in the hotter clinkering zone, more expensive, higher temperature basic refractories (magnesia, doloma, spinel or magnesia-chrome) have to be installed. Regular monitoring of the outer shell for hot spots

CEMENT KILNS

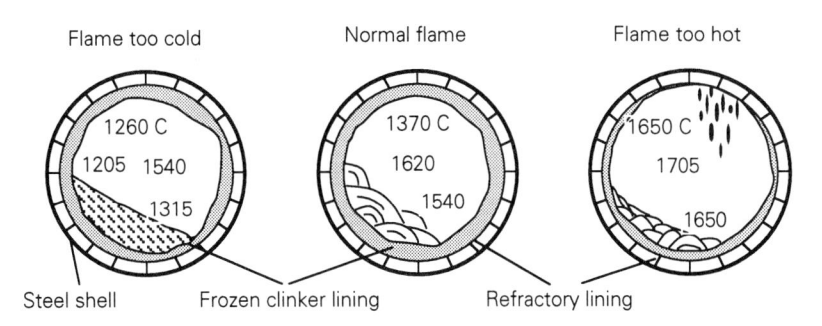

Figure 2.9 The effect of flame temperature on the refractory linings (redrawn from Peray, 1986, p. 151)

is essential in order to detect any refractory failure as this would cause the steel shell to distort and warp to such an extent that replacement of that section of the steel shell becomes necessary. The damage can be minimized if the kiln is shut down and lining repairs made as soon as refractory failure is detected.

In the flame region, some reaction between the clinker and the refractories is desirable in order to form a coating which protects the refractory bricks from abrasion; however excessive build-up of the coating has to be avoided. This is done by careful control of the flame temperature. If the flame is too hot the protective coating will melt and the refractories can be damaged; if it is too cold excessive build-up of the coating and incomplete reaction of the cement clinker can occur (as shown in Figure 2.9).

The nature of the refractory also plays a part in the satisfactory build-up of a protective coating. Magnesia-chrome refractories have excellent properties, but the use of these refractories is now being phased out because of the problems of the disposal of used refractories which have been found to contain unacceptably high levels of water-soluble Cr^{6+} which potentially could contaminate water supplies. Spinel-based refractories (magnesium aluminium oxide) are now being used, but it is difficult to form satisfactory protective coatings on this material and the ideal refractory for the clinkering part of the kiln has yet to be found.

The life of the refractory has been found to be critically dependent on the number of times the kiln is shut down. The more shut-downs, the more the damage to the refractory. The major cause of refractory loss during shut-down is spalling due to thermal shock which occurs as the refractory is cooled. Ideally the kilns should be kept running continuously, as this would result in the refractories being kept at constant temperature, but if shut-downs have to occur then it is essential that the cooling is slow and uniform so as to minimize the thermal shock to the refractory lining.

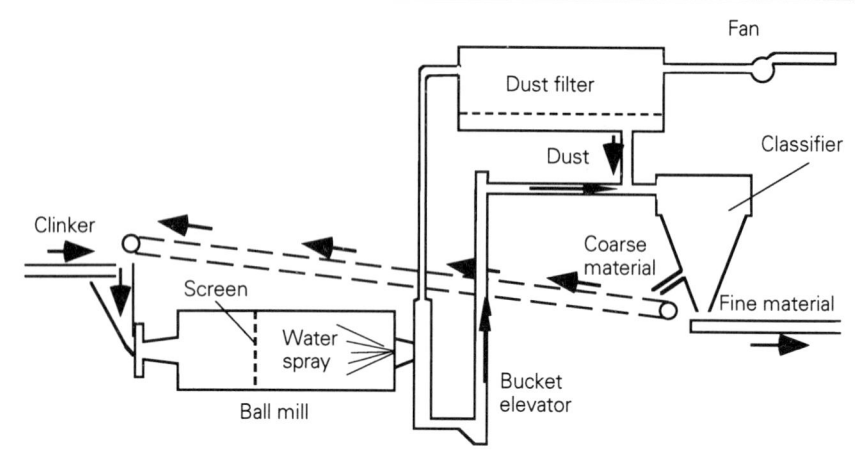

Figure 2.10 Clinker grinding set up (redrawn from Kerton *et al.*)

2.1.6 Grinding of cement clinker

Once the clinker is cooled it is ground in a ball mill to a powder to produce cement. The diameter of the ball mill is typically about 4 metres. The layout is shown diagrammatically in Figure 2.10.

In the ball milling process, some 97–99% of the energy supplied to the mill is converted into heat. Hydrated calcium sulphate (gypsum) is added to the cement at this stage in order to control the early rate of hydration of the cement. If the gypsum is heated to above 180 °C it partially dehydrates to form plaster of Paris which would cause other problems during the setting of the cement. It is therefore essential that the mill be cooled but the large diameter of the mills makes external water cooling inefficient, and it is necessary to cool the ground cement clinker and the added gypsum with a water spray. It should be appreciated that the injected water spray has to be very carefully monitored; only enough has to be used to cool the powdered cement by complete evaporation of the water to produce a dry product. If this is not the case, then the fine cement particles adhere to each other and lumps are formed during the storage of the cement in the silos (Kerton and Murray, 1983).

2.2 RAW MATERIALS

The exact nature of the raw materials used varies widely from one country to another and is dependent on the nature of local mineral deposits. The calcareous materials ($CaCO_3$) may be chalk (a soft limestone of organic origin, being mainly the shells of marine organisms) or limestone (which is usually ancient sedimentary limestone which has undergone geological changes). The argillaceous materials (aluminosilicates) include clays, shales and marls, mudstone, slate, schist, and some volcanic rocks and ash. The series clay, shale, mudstone and slate are different stages in rock formation. They are chemically similar, but their physical properties are different and this has to be taken into account in the pregrinding process (Pollit, 1964).

Artificial materials are also sometimes added. Blast furnace slag is one of the most important of these. Other materials such as sand, waste bauxite, and iron oxide may be used in order to adjust the composition of the mix.

2.3 PHASES IN PORTLAND CEMENT CLINKER

After the most commonly used raw materials have been reacted in the cement kiln, four major phases in Portland cement are formed during the clinkering, as shown in Table 2.2.

The setting of Portland cement is due to the hydration of these phases and the raw materials (lime and clay) must be mixed in the correct proportions so that the final composition of the clinker contains these phases.

2.3.1 Ternary phase diagram for Portland cement clinker

For simplicity, cement composition will be initially expressed in terms of **C₃S, C₂S** and **C₃A**. In this case, the composition of the cement must lie in the triangle bounded by **C₃S, C₂S** and **C₃A**, shown in the ternary **C-S-A** phase diagram in Figure 2.11.

Table 2.2 Major phases present in Portland cement clinker

Real phases in clinker	Idealized chemical composition		Abbreviation
Alite	3CaO.SiO₂	(tricalcium silicate)	(**C₃S**)
Belite	2CaO.SiO₂	(dicalcium silicate)	(**C₂S**)
C₃A-alkali solid solution	3CaO.Al₂O₃	(tricalcium aluminate)	(**C₃A**)
Ferrite phase solid solution	4CaO.Al₂O₃.Fe₂O₃	(calcium alumino ferrite)	(**C₄AF**)

Notes:
1. The phases present in the cement clinker can contain other compounds in solid solution with the ideal composition.
2. In order to simplify the compositions and equations used in cements, the following abbreviations are made: **C** = CaO, **S** = SiO₂, **A** = Al₂O₃, **F** = Fe₂O₃.
3. The presence of **C₄AF** is due to iron impurities which are normally present in the raw materials.

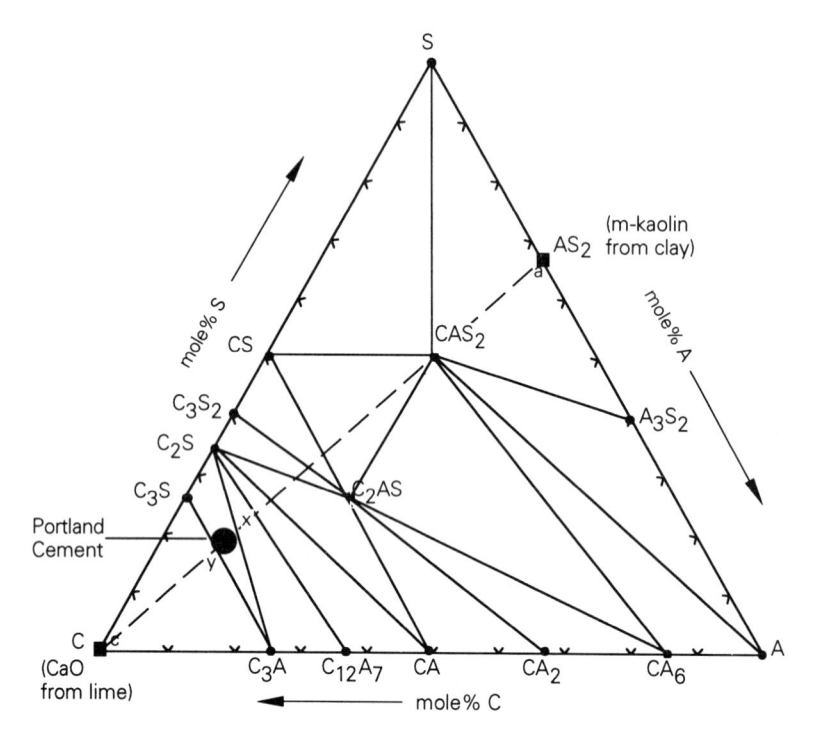

Figure 2.11 CaO-SiO$_2$-Al$_2$O$_3$ phase diagram showing compatibility triangles

The relative amounts of the raw materials (lime and clay (kaolin)) needed to form clinker with the desired composition can be found from Figure 2.11.

The composition of m-kaolin (Al$_2$O$_3$.2SiO$_2$) which is formed when kaolin (Al$_2$O$_3$.2SiO$_2$.2H$_2$O) is heated to about 550 °C and loses structural water, is shown by point a on the diagram. Calcium oxide (CaO) is formed by the decomposition of limestone (CaCO$_3$), and is shown by point c. The compositions that can be made by mixing calcium oxide and m-kaolin will all lie on the line joining these two compositions, ac. Because the composition of the Portland cement clinker must lie within the triangle joining **C$_3$S, C$_2$S and C$_3$A**. The maximum mole fraction of m-kaolin to calcium oxide is given by the ratio of the lengths cx/ac (0.31), and the minimum fraction of m-kaolin to calcium oxide is the ratio cy/ac (0.26).

In order to form the cement clinker at a reasonable rate, it is necessary to have partial melting of the clinker. For partial melting to occur, a minimum temperature of 1455 °C has to be achieved.

This is shown in the more detailed part of the **C-S-A** phase diagram in Figure 2.12.

PHASES IN PORTLAND CEMENT CLINKER

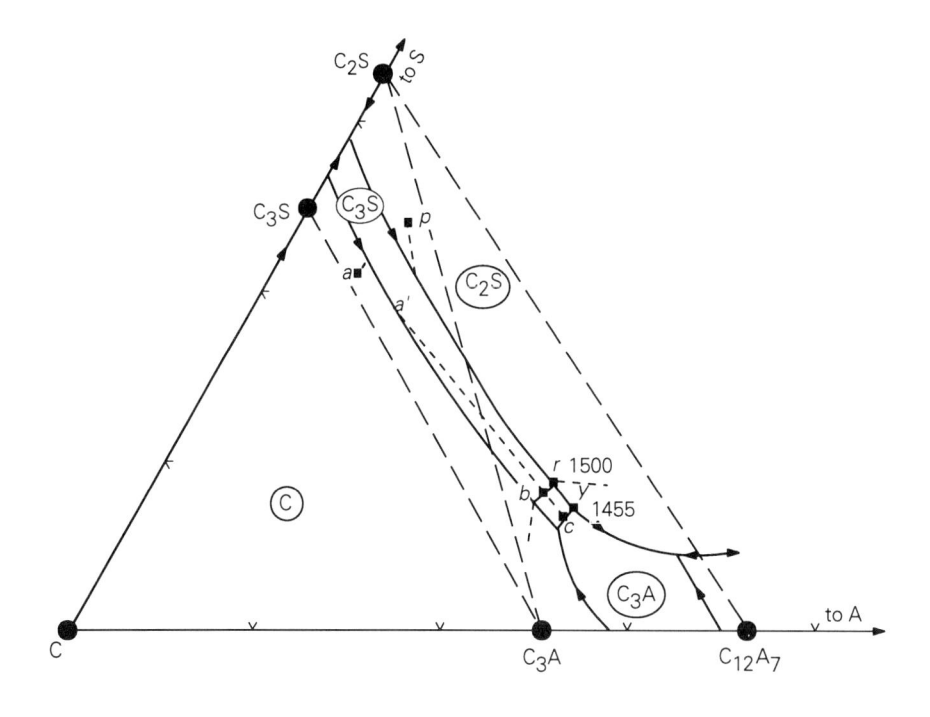

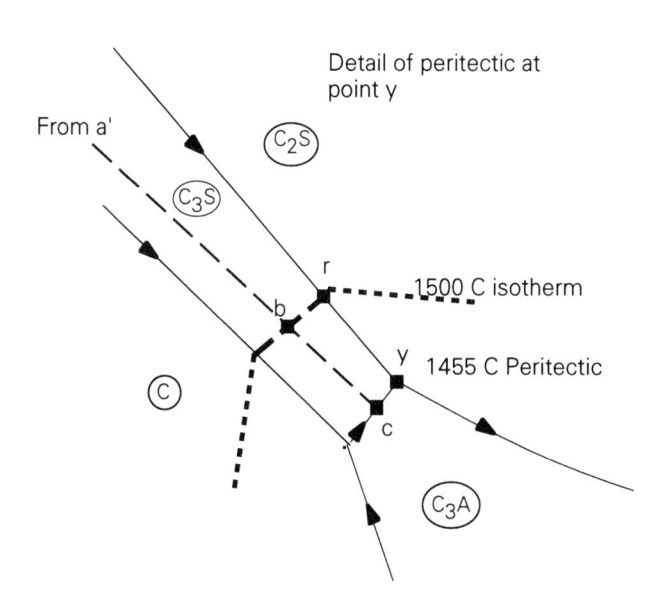

Figure 2.12 Part of the system CaO-SiO_2-Al_2O_3

For all compositions within the triangle bounded by C_3S, C_2S and C_3A, the first liquid will be formed when the materials are heated to 1455 °C, and the composition of the liquid is given by point y.

The phase diagram can also be used to determine how the relative proportions of the phases present in the clinker and the microstructure of the clinker are influenced by the relative proportions of the raw materials used (Lea, 1970).

The relative proportions of the final phases can be found by the centre-of-gravity rule. Some idea of the microstructure can be gained by the determination of the sequence in which the solid phases precipitate out of a liquid as it solidifies.

If the composition of the cement is given by point a in Figure 2.12, the changes in liquid composition which would take place during the solidification of a liquid of this composition can be determined as follows. In this case, point a lies in the stability field of C, and therefore C will be the first solid to precipitate out of the liquid. When the liquid composition reaches the boundary between the stability fields of C and C_3S, C_3S will start to precipitate out, while C redissolves in the liquid. When all the C has redissolved (point a′) the liquid composition will move along the line a′c, during which time more solid C_3S is precipitating from the liquid. At point c, which is on the boundary between the fields of stability of C_3S and C_3A, solid C_3A will start to form, and the liquid composition will then move to the peritectic at y, where the remaining liquid will freeze to form C_3S, C_3A and C_2S at 1455 °C. Thus the minimum temperature for the production of a liquid phase is 1455 °C. In practice the clinker is heated to 1500 °C, at which stage the phases present are solid C_3S and a liquid of composition b. The cooling sequence from b is as described above.

Clinker, which is relatively rich in calcia, will consist of primary precipitates of C_3S and C_3A embedded in a fine ternary mixture of C_3S, C_2S and C_3A.

In practice not all of the liquid crystallizes and there will be some glass formation, particularly if the clinker is cooled rapidly.

If the cement composition is richer in silica (e.g. at point p), then the cooling route is different, C_2S will precipitate out first, and then both C_2S and C_3S will form simultaneously until the liquid composition reaches y, at which point it will freeze to form a fine mixture of C_3S, C_2S and C_3A. If the clinker of this cement composition is formed at 1500 °C there will be present at this temperature solid C_2S and C_3S and a liquid of composition r.

Increasing the silica content of the clinker results in a change in the microstructure of the clinker which will contain particles of C_2S and C_3S embedded in a microcrystalline matrix of C_3S, C_2S and C_3A or a glass.

Further increase in the silica content will result in a change of the

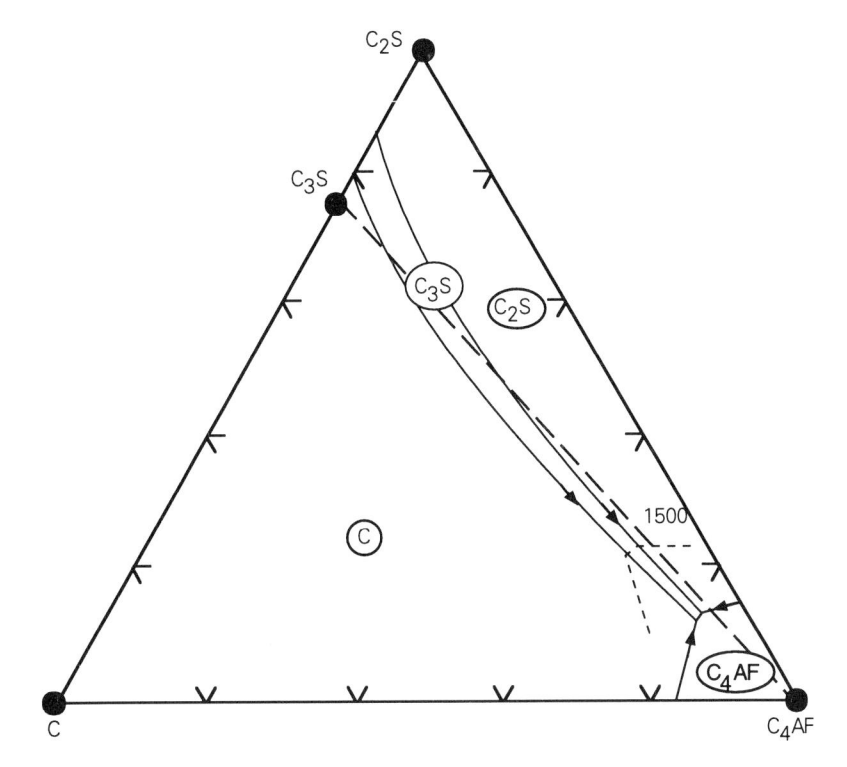

Figure 2.13 Part of the CaO-SiO$_2$-Fe$_2$O$_3$ phase diagram

phases present in the cement to **C$_3$S, C$_2$S** and **C$_{12}$A$_7$**. (The presence of **C$_{12}$A$_7$** is undesirable as it hydrates too rapidly.)

If there are iron oxides present in the reactants, the Fe$_2$O$_3$ tends to replace the Al$_2$O$_3$. If the ratio Al$_2$O$_3$/Fe$_2$O$_3$ = 0.64 there will be no **C$_3$A** formed in the clinker. The product of reaction is then **C$_4$AF**. The phase diagram (Figure 2.13) shows that at 1500 °C solid **C$_3$S** and **C$_2$S** and liquid are present. More **C$_2$S** and **C$_4$AF** are produced during cooling.

The presence of iron oxide alters the composition of the phases, in that there will be an increase in **C$_4$AF** with a corresponding decrease in **C$_3$A**. In practice, most Portland cements lie between these two extremes, and both **C$_3$A** and **C$_4$AF** are present in the clinker.

2.3.2 Quaternary phase diagram for Portland cement clinker

On the quaternary diagram (Figure 2.14) the range of Portland cement compositions lie in the tetrahedron bounded by **C$_3$S, C$_2$S, C$_3$A** and **C$_4$AF**. Within this volume is a quaternary invariant point at 1338 °C, which occurs

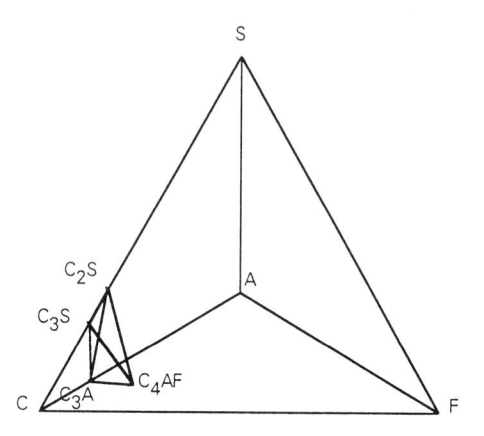

Figure 2.14 Al_2O_3-CaO-SiO_2-Fe_2O_3 phase diagram

at the composition (in wt.%) **C** = 54.8, **A** = 22.7, **F** = 15.5 and **S** = 6.0 (Roy, 1983).

2.3.3 Reactions that occur in the kiln

The completeness of reactions which take place in the kiln is dependent on the mixing and particle size of the reactants. It is essential that the raw materials be finely ground and well mixed. Quartz, which is normally an impurity in the clay, is particularly troublesome since it is the most refractory of the raw materials and it must be finely ground to be completely reacted during the formation of the clinker.

Typical amounts (mass%) of reactants used to form Portland cement, and the amounts of the phases present in the clinker are shown in Table 2.3.

C₃S and **C₂S** are the most important components in the cement clinker, as they are responsible for the long term strength of the cement which develops after the cement is mixed with water. The sequence of reactions that take place in a dry process kiln is listed in Table 2.4.

The maximum temperature, which is reached in the burning or clinkering zone is normally between 1300 and 1500 °C. This is chosen to produce a degree of melting just sufficient to cause the material to cohere into small balls (nodules or lumps) of clinker. The amount of liquid present is between 20 and 30%. Higher temperatures are avoided because there would be problems in the flow of material through the kiln as well as promoting attack on the refractory linings. More importantly, higher temperatures would result in the formation of large crystals of the reaction products in the clinker which would lower their subsequent reactivity with

PHASES IN PORTLAND CEMENT CLINKER

Table 2.3 Typical amounts of reactants and products (mass%) in Portland cement clinker

Reactants

	CaO	60–67	from $CaCO_3$
	SiO_2	17–25	from clay
	Al_2O_3	3–8	from clay
	Fe_2O_3	0.5–6	from clay
	Na_2O, K_2O	0.5–1.3	from clay or impurities
	MgO	<6	impurities
	SO_3	1–3	impurities

Products

	alite (~ **C_3S**)	~45
	belite (~ **C_2S**)	~25
	tricalcium aluminate (~ **C_3A**)	~10
	calcium alumino ferrite (~ **C_4AF**)	~10
	other phases	~10

Table 2.4 Chemical and physical changes that take place in the production of Portland cement clinker

~500°C	Loss of structural water from the clay minerals
~900°C	Chemical reaction of m-kaolin
~900°C	Decomposition of $CaCO_3$
900–1200°C	Reaction between CaO and aluminosilicates
1250–1280°C	Beginning of liquid formation
>1280°C	Further liquid formation and completion of the formation of cement compounds

Note that the liquid formation occurs at a lower temperature than that which was predicted from the phase diagrams. This is due to the fluxing action of the minor impurities present in the raw materials.

water. This, in turn, would reduce the rate of setting of cement which is a result of the hydration of the reaction products in the clinker.

2.3.4 Calculation of the composition of Portland cements

The potential composition of Portland cement clinker can be calculated from equations based on the Bogue calculation (Mindess, 1983). The percentages of the main compounds are shown below. The terms in brackets are the wt.% of the oxide in the cement formulation.

(a) for $A/F \geq 0.64$

$C_3S = 4.07\ (C) - 7.60\ (S) - 6.72\ (A) - 1.43\ (F) - 2.85\ (\bar{S})$

$C_2S = 2.87\ (S) - 0.75\ (C_3S)$

$C_3A = 2.65\ (A) - 1.69\ (F)$

$C_4AF = 3.04\ (F)$

where $\bar{S} = SO_3$

(b) for $A/F < 0.64$

$C_3S = 4.07\ (C) - 7.60\ (S) - 4.48\ (A) - 2.86\ (F) - 2.85\ (\bar{S})$

$C_2S = 2.87\ (S) - 0.75\ (C_3S)$

$C_3A = 0$

$C_4AF = 2.10\ (A) + 1.70\ (F)$

The above equations assume that the reactions proceed to completion, but in industrial practice there are restrictions on both the burning and reaction times which means that there are deviations from the amounts calculated from the above formulae.

Minor amounts of other phases are found, the most important of which is unreacted CaO. If it is known that there is free lime present, then this should be subtracted from the total CaO content before the calculation is carried out. Negative values for C_2S content show that free lime must be present (Kohlhaas, 1983, p. 134).

Quantitative X-ray diffraction techniques have been used to measure the amounts of the crystalline phases present in cement clinker. These have shown that the inaccurate results given by the Bogue calculations are not only due to the non-equilibrium conditions but are also the result of solid solutions formed by the phases with other ions such as magnesium, potassium, sodium, iron etc. by the four major phases. To overcome this problem, Taylor (1989) describes a procedure which is essentially a solution of four linear simultaneous equations which produces more accurate results.

Other chemical composition parameters are useful when manufacturing processes are considered (Roy, 1981); these are the silica and alumina moduli and the lime saturation factor where:

Silica modulus (SM) $\quad\quad\quad S / (A + F)$
Alumina modulus (AM) $\quad\quad A / F$
Lime saturation factor (LSF)%100 °C / $(2.8\ S + 1.18\ A + 0.65\ F)$

These factors can be used to determine how much liquid phase is present during the burning process and this, in turn, will affect the rate of reaction between the reactants (Scrivener, 1989).

(a) Silica modulus

The amount of liquid phase is dependent on the value of this ratio. Typical values of the silica modulus (SM) are between 2.3 and 2.5. If SM is too high, then the amount of liquid phase produced is low, which results in not all the material being converted into nodules of clinker. The remaining unmelted dusty material tends to clog the kiln and also be incompletely reacted.

(b) Alumina modulus

The alumina modulus (AM) affects the temperature at which melting commences. Typical values are about 2. The lowest temperature at which liquid is formed occurs at AM = 1.6, and this is the optimum for the formation of clinker material and nodulization.

(c) Lime saturation factor

Complete reaction of the calcium oxide in the mix to form compounds can be expected if the lime saturation factor is < 100%. If the value is > 100% then there will always be some free lime left in the clinker. Typically the LSF is 92–96%.

2.3.5 Effect of impurities and cooling rate on the clinker

The products at the clinkering temperature are normally **C_3S** and **C_2S**, with a liquid containing CaO together with Al_2O_3, Fe_2O_3 and MgO, but relatively little SiO_2. The ferrite and aluminate phases form on the freezing of the liquid, together with magnesium oxide. In earlier studies it was considered that rapid cooling of the clinker resulted in glass formation, with the glass content varying from 5 to 25%. However this is now disputed (Taylor, 1990, p. 85). The identification of the 'glass' phase was done by the use of optical microscopy, which is unable to detect finely crystalline material or distinguish between cubic crystals and glass. X-ray diffraction, Scanning electron microscopy and X-ray microanalysis studies all support the formation of a microcrystalline phase. (Glass has been observed in clinkers that have been rapidly quenched in water, but this does not occur in the normal production of the clinker.)

The magnesium oxide impurity has the advantage of acting as a flux during the formation of the clinker and increases the rate of reaction of the components, but the presence of magnesium oxide in the clinker can cause problems after the hydration of the cement as it could cause the cement to expand after setting has taken place. The maximum amount of MgO that can be present is prescribed by standards. (The maximum limit varies with the different standards.)

Even though alkalis (potassium and sodium) are usually only present in small amounts, their presence can cause problems in concrete; they react with certain aggregates and the products of this alkali-aggregate reaction can cause disintegration of the set concrete.

The cooling rate affects the physical properties of the cement in four main ways.

- Firstly, cements made from slowly cooled clinker often show a 'flash set', even when mixed with water in the presence of gypsum which has been added to the cement to prevent this happening. (A 'flash set' is the very rapid setting of cement accompanied by high evolution of heat, which makes the cement very hard to work and produces a low strength cement. It is linked with the presence of C_3A.)
- Secondly, if the clinker has a relatively high MgO content (2.5–5.0%), slow cooling results in the formation of a cement which gradually expands after hardening (probably due to the hydration of the MgO). This is because slow cooling permits the precipitation of MgO crystals, whereas rapid cooling results in the MgO remaining in solid solution in the aluminate or ferrite phases.
- Thirdly, slow cooling can result in the decomposition of the C_3S phase to form C_2S and free C (Taylor, 1990, p. 88). This happens in the region between 1250 and 1100 °C (see the CaO-SiO_2 phase diagram in Figure 2.15). This has recently been questioned by Johansen (1989, p. 43) who considers that the decomposition of the C_3S into free lime and C_2S is a very slow process below 1250 °C and is of no consequence in the production of the clinker. Glasser (1983) discusses the effects of cooling rate and the catalytic effect of impurities on the decomposition of C_3S in detail.
- Fourthly, rapid cooling of the clinker results in the formation of cracks within the clinker nodule due to thermal shock. The cracked clinker is more easily ground to a fine cement powder.

All the above points indicate that the clinker should be rapidly cooled, since this results in:

- the C_3A phase reacting more slowly with water when it is fine grained and mixed with the C_4AF phase;
- the MgO remains in solid solution;
- the possible decomposition of the C_3S phase is avoided;
- and better grindability of the clinker.

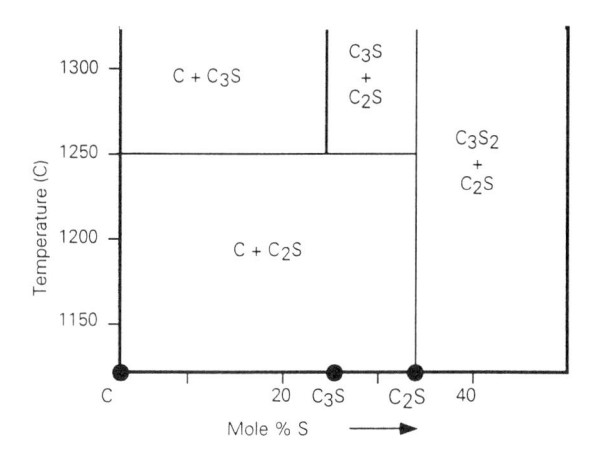

Figure 2.15 CaO-SiO$_2$ phase diagram

2.3.6 Microstructure of the clinker

The microstructure of the clinker depends on the composition of the cement and the cooling rate. However, in general, it consists of large particles of **C$_3$S** and **C$_2$S** immersed in a matrix of a microcrystalline solid (or a glass) in which are embedded smaller particles of **C$_2$S, C$_3$S, C$_3$A** etc. The alite (**C$_3$S**) crystals are hexagonal in shape, whereas the belite (**C$_2$S**) tend to be rounded (Johansen, 1989, p. 27). As shown in Figure 2.16, the alite crystals are large and have straight boundaries whereas the smaller belite crystals have curved boundaries (Kohlhaas, 1983, p. 135). The crystals in the clinker are often cracked due to thermal stresses during the cooling of the clinker (the cracking of the clinker due to thermal stress is an advantage as it aids the grinding process). A typical microstructure of cement clinker is shown in Figure 2.16.

QUESTIONS

1. What are the advantages of using a rotary dry-process kiln with a precalciner over a wet process kiln for the production of Portland cement clinker?
2. Why must the flame temperature in the cement kiln be carefully controlled?
3. What raw materials are used for the production of Portland cement? What are the four major crystalline phases present in Portland cement clinker?

PRODUCTION OF PORTLAND CEMENT

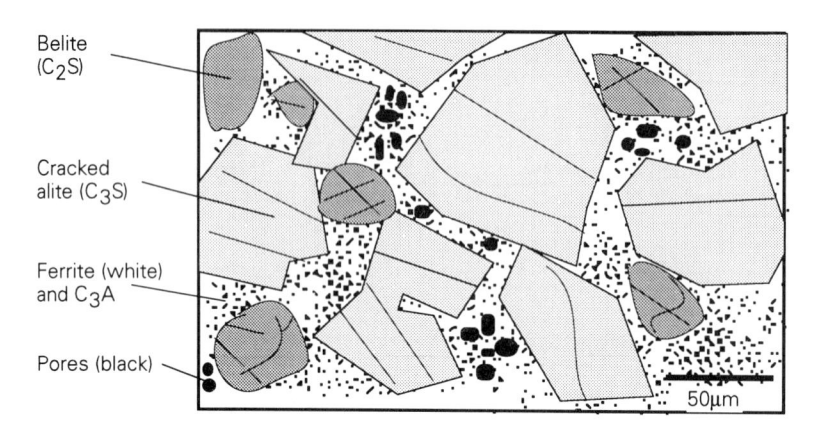

Figure 2.16 Microstructure of a commercial clinker. The alite crystals are light grey and the belite are the darker rounded crystals. The phases in between the alite and belite are ferrite (white) and **C₃A** (darker). The pores are black (redrawn from a micrograph from Scrivener, 1989)

4. What are the maximum and minimum ratios of silica:calcia (by weight) that can be used to produce Portland cement? What happens if the composition lies outside this range? (Molecular weight SiO_2 = 60, CaO = 56.)
5. Why is it necessary to partially melt the reactants in the production of cement clinker?
6. The chemical composition (mass %) of a Portland cement is SiO_2 (22.8), Al_2O_3 (3.6), Fe_2O_3 (4.5), CaO (64.5) and remainder (4.6).
   (a) Calculate the theoretical phase composition of the cement.
   (b) Why will the actual phase composition differ from that calculated?
   (c) What would be the effect of increasing the Fe_2O_3 content to >5.6 mass %?
   (d) What is likely to be present in the 'remainder'?
   (e) Calculate the Silica Modulus, Alumina Modulus and Lime Saturation Factor, and comment on the significance of the results.
7. Discuss the effects of fast and slow cooling rates of Portland cement clinker.

References

Bulavin, I., Makarov, I., Rapoport, A. *et al.* (1986) *Heat Processes in Glass and Silicate Technology* (translated by V. Afanasyev), MIR Publishers, Moscow.

Glasser, F. P. (1983) Reactions occurring during cement making. In P. Barnes (ed.), *Structure and Performance of Cements*, Applied Science, London, pp. 69–108.

REFERENCES

Johansen, V. (1989) Cement production and cement quality. In J. P. Skalny (ed.), *Materials Science of Concrete I*, The American Ceramic Society, Inc., Westerville, OH.

Kelly, K. E. and Beahler, C. C. (1993) Burning hazardous waste in cement kilns: a study of economics, offsite concentrations, and health effects in Midlothian, Texas. IEEE Cement Industry Conference XXXV, 363–94.

Kerton, C. P. and Murray, R. J. (1983) Portland cement production. In P. Barnes (ed.), *Structure and Performance of Cements*, Applied Science, London, pp. 205–36.

Kohlhaas, B. (1983) *Cement Engineers' Handbook*, 4th English edn, Bauverlag, Weisbaden.

Lea, F. M. (1970) *The Chemistry of Cement and Concrete*, 3rd edn, Edward Arnold, London, Ch. 7.

Mindess, S. (1983) Concrete materials. *Journal of Materials Education*, 4, 984–1046.

Morton, E. (1991) Burning waste fuels in cement kilns. *IEEE Cement Industry Conference XXXIII*, 155–70.

Osbaeck, B. (1994) The use of secondary products in the cement and concrete industry. In P. W. Brown (ed.), *Cement Manufacture and Use*, American Society of Civil Engineers, New York, pp. 107–21.

Peray, K. E. (1986) *The Rotary Cement Kiln*, 2nd edn, Chemical Publishing Co., New York.

Pollitt, H. W. W. (1964) Raw materials and processes for Portland cement manufacture. In H. F. W. Taylor (ed.), *The Chemistry of Cements, Vol 1*, Academic Press, London, pp. 27–48.

Redeker, A. E. (1994) The economics of waste fuel burning. *IEEE Cement Industry Conference XXXVI*, 185–98.

Roy, D. M. (1981) Portland cement: constitution and processing. part 1: cement manufacture. *Journal of Educational Modules for Materials Science and Engineering*, 3, 626–47.

Roy, D. M. (1983) Portland cement: constitution and processing. Part II Cement constitution and kiln reactions. *Journal of Materials Education*, 5 (3), 465–90.

Scrivener, K. L. (1989) The microstructure of concrete,. In J. P. Skalny (ed.), *Materials Science of Concrete I*, The American Ceramic Society, Inc., Westerville, OH, p. 127.

Taylor, H. F. W. (1989) Modification of the Bogue Calculation. *Advances in Cement Research*, 2 (6), 73–7.

Taylor, H. F. W. (1990) *Cement Chemistry*, Academic Press, London.

3 | Hydration of cement – setting reactions

3.1 HYDRATION REACTIONS OF INDIVIDUAL COMPONENTS IN CEMENT CLINKER

The cement powder that is produced by ball milling the clinker consists of fine multiphase particles ranging in size from about 1 to 100 mm. The hydration reactions that occur when the cement is mixed with water are complex and interdependent. In order to simplify the investigations most studies have been carried out on the hydration reactions of the individual phases in the cement, but it should be noted that the phases present in the cement clinker are not chemically pure and extensive solid solubility can occur. For example, tricalcium silicate (C_3S) formed in the clinker, can contain up to 2 wt% MgO as well as some Al_2O_3, Fe_2O_3, TiO_2 and other oxides in solid solution. Dicalcium silicate can form solid solutions with Na_2O, K_2O, Al_2O_3, Fe_2O_3 and F. Tricalcium aluminate can incorporate over 5 wt% Na_2O and K_2O in its structure and the composition of the ferrite phase is variable and can contain MgO (Kohlhaas, 1983).

3.1.1 Hydration of C_3S (tricalcium silicate)

When mixed with water, the C_3S phase hydrates to form crystalline $Ca(OH)_2$ (variously known as **CH**, calcium hydroxide, lime or portlandite) and a gel of hydrated $CaO\text{-}SiO_2$ (**CSH**). The overall reaction can be shown as (Jawed et al., 1983):

$$2C_3S + 7H \rightarrow C_3S_2H_4 + 3CH \quad \Delta H = -114 \text{ kJ/mol}$$

This does not exactly describe the reaction because the hydrated calcium silicate is normally non-stoichiometric and therefore is usually written as **CSH**. This **CSH** gel is microcrystalline or non-crystalline.

HYDRATION REACTIONS OF INDIVIDUAL COMPONENTS

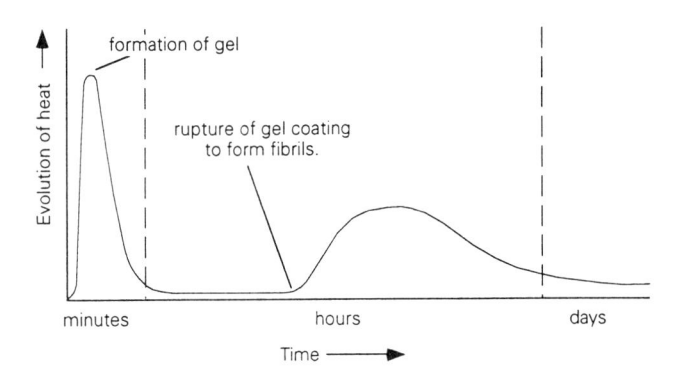

Figure 3.1 Schematic representation of the rate of heat evolution from C_3S

The reaction occurs by the surface of C_3S dissolving in water to form a very dilute solution. The products of the reaction have very low solubility and form a semipermeable membrane on the solid surface of C_3S. The initial reaction is rapid, but the rate slows down as the C_3S becomes covered with the reaction products. During this time the small calcium ions (Ca^{2+}) diffuse out from the C_3S through the semipermeable membrane into the water. The silicate ions are too large to pass through the membrane and remain trapped close to the surface of the C_3S. The H_2O and H_3O^+ from the water pass through the membrane towards the solid surface of the C_3S. The **C:S** ratio of the initial coating is near 3. After a few hours the coating ruptures due to the increase in osmotic pressure caused by the diffusion of the ions through the membrane. This finally results in the formation of another **CSH** gel which has a **C:S** ratio of 1.5 or less. When the membrane ruptures, the silicate ions that were trapped by the membrane are released and can react with the calcium ion rich water to form tubular fibrils. The kinetics of the reaction can be followed by measuring the rate at which heat is evolved during the hydration, as shown in Figure 3.1 (redrawn from Jawed *et al.*, 1983).

The composition of the **CSH** gel changes with time, and many different gels have been identified. A detailed description of the formation of the gels, structural models and structural relationships is given by Taylor (1990, Ch.3).

The calcium hydroxide produced by the reaction makes the cement alkaline (pH ~12.5).

A diagrammatic representation of the initial reactions is shown in Figure 3.2.

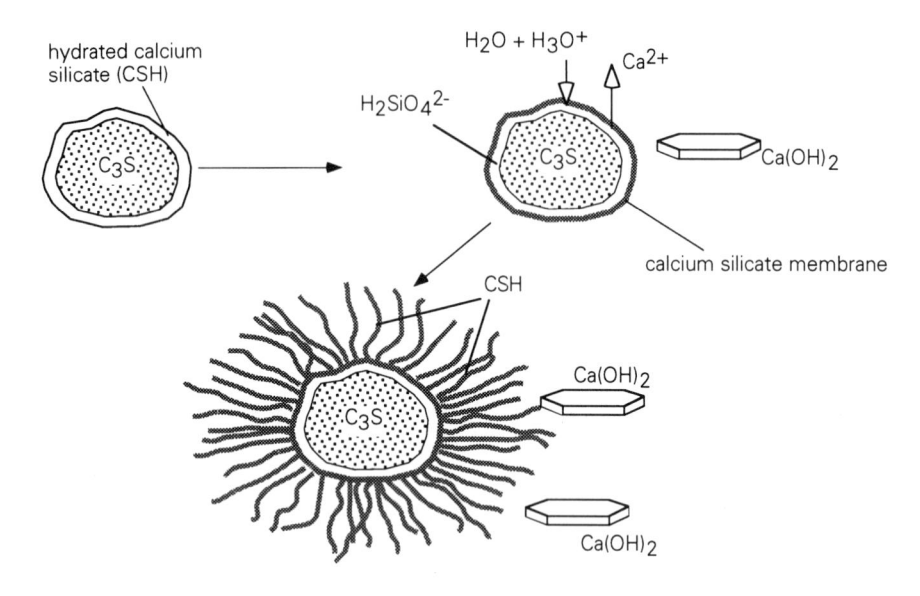

Figure 3.2 Representation of the initial hydration reactions of C_3S

3.1.2 Hydration of C_2S (dicalcium silicate)

Dicalcium silicate occurs in four different crystalline forms namely α, α', β and γ. It is the β form that normally exists in Portland cement. β-C_2S reacts very slowly with water to form mainly **CSH** gel (no, or very little, **CH** is formed). Again, the gel forms on the surface of the solid, and the thickening gel layer slows down the rate of hydration. The rate of hydration is about 20 times slower than the hydration of the C_3S, after four years about 15% of the C_2S remains unreacted. However, the gel formed is similar to that produced by the C_3S.

The overall reaction can be represented by (Jawed *et al.*, 1983):

$$2C_2S + 5H \rightarrow C_3S_2H_4 + CH \quad \Delta H = -43 \text{ kJ/mol}$$

3.1.3 Hydration of C_3A (tricalcium aluminate)

(a) With gypsum additions (sufficient to react with all the C_3A)

Gypsum (calcium sulphate dihydrate, $CaSO_4.2H_2O$ or $C\bar{S}H_2$ where $\bar{S} = SO_3$) is normally added to the cement to react with the C_3A. This is done to help control the hydration rate of C_3A and it does so by reacting with the C_3A to first form a hydrated calcium aluminosulphate (ettringite) which is only stable in the presence of an ample supply of gypsum. If the molar ratio of added gypsum to C_3A is 1, then the reaction to form

HYDRATION REACTIONS OF INDIVIDUAL COMPONENTS

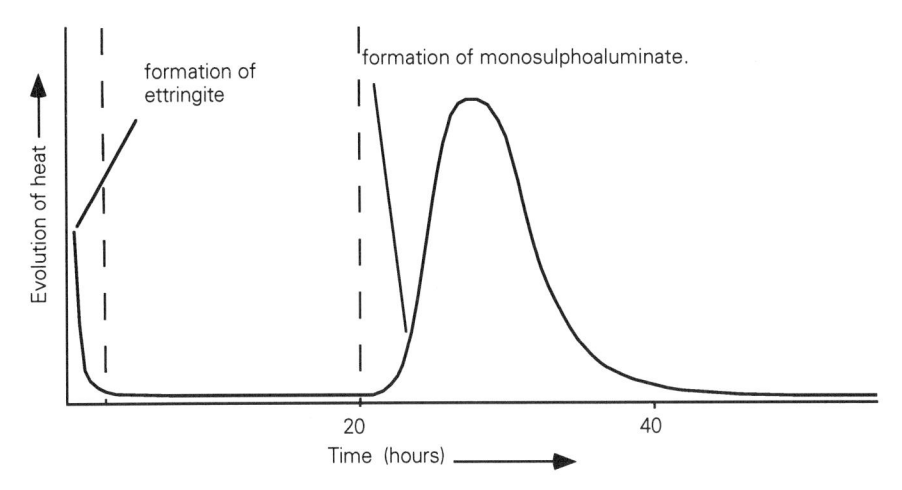

Figure 3.3 Effect of ettringite formation on the evolution of heat by the hydration of C_3A

ettringite lowers the gypsum concentration and the ettringite becomes unstable and converts to a monosulphoaluminate hydrate (monosulphate), as shown below (Jawed et al., 1983).

$$C_3A + 3C\bar{S}H_2 + 26H \rightarrow C_6A\bar{S}_3H_{32} \quad \text{ettringite}$$

$$C_6A\bar{S}H_{32} + 2C_3A + 4H \rightarrow 3C_4A\bar{S}H_{12} \quad \text{monosulphoaluminate hydrate}$$

The formation of ettringite results in a retardation of the hydration of the C_3A. The heat of reaction (–362 kJ/mol of C_3A reacted) is also liberated over a longer period of time and thus does not cause a large rise in temperature, as shown in Figure 3.3.

Note: There are several ways of writing the formulae for ettringite and monosulphoaluminate as shown in Table 3.1 (Young, 1981).

(b) In the absence of gypsum

In the absence of gypsum, C_3A reacts very rapidly with water. Crystalline phases form in a few minutes and increase in size and quantity very rapidly (Jawed et al., 1983). The reaction is:

$$2C_3A + 21H \rightarrow C_4AH_{13} + C_2AH_8 \quad \text{hexagonal hydrates} \quad \Delta H = -340 \text{ kJ/mol}$$

These hydrates are members of the AF_m group and are structurally related to monosulphoaluminate. These phases are metastable and eventually form hydrogarnet (C_3AH_6).

Table 3.1 Alternative formulae for ettringite and monosulphoaluminate hydrate

Ettringite	Monosulphoaluminate hydrate
$C_6A\bar{S}_9H_{32}$	$C_4A\bar{S}H_{12}$
$C_3A.3C\bar{S}_3.H_{32}$	$C_3A.C\bar{S}.H_{12}$
AF_t	AF_m

AF_t refers to a general group of ettringite phases, Al_2O_3-Fe_2O_3-trisulphate, in which the Al can be partially or fully replaced by Fe, and the sulphate by other anions. These phases have a pseudo-hexagonal structure and are less dense than the AF_m monosulphate aluminate phases.

AF_m is the more general group of monosulphoaluminate phases, Al_2O_3-Fe_2O_3-**m**onosulphate, in which partial or complete substitution of Al by Fe can occur. The monosulphoaluminate phases are hexagonal and are more dense than the AF_t ettringite phases.

$$C_4AH_{13} + C_2AH_8 \rightarrow 9H + 2C_3AH_6 \text{ hydrogarnet} \quad \Delta H = -70 \text{ kJ/mol}$$

The rate at which the metastable phases change to the stable cubic phase increases with increasing temperature. The large amount of heat that is liberated in the formation of the metastable phases is sufficient to raise the temperature by several tens of degrees which accelerates conversion of the metastable phases to the stable hydrogarnet phase.

(c) Intermediate amounts of gypsum

If there is insufficient gypsum to react with all the C_3A, there will be some unreacted tricalcium aluminate left when all the ettringite has converted to the monosulphoaluminate. In this case, solid solutions are formed with composition in between $C_4A\bar{S}H_{12}$ and C_4AH_{13}. The reaction can then be given by (Young, 1981):

$$C_4A\bar{S}H_{12} + C_3A + CH + 12H \rightarrow 2[C_4A\bar{S},H)H_{12}]$$

(d) Excess gypsum

The presence of excess gypsum, results in the ettringite being stable and the reaction between C_3A and sulphate will continue after the cement has set. The continued reaction can result in the cracking of the set cement (sulphate attack).

(e) Summary of the hydration products of C_3A

A summary of the hydration products of C_3A is shown in Table 3.2.

Table 3.2 The effects of gypsum on the products of hydration of C_3A

Molar ratio (gypsum) $(C\bar{S}H_2)/C_3A$	Stable products of hydration
3.0	ettringite
3.0–1.0	ettringite + monosulphoaluminate
1.0	monosulphoaluminate
<1.0	solid solution of $C_4A\bar{S}H_{12} + C_4AH_{13}$
0	hydrogarnet (C_3AH_6)

3.1.4 Hydration of C_4AF

C_4AF reacts with water in a manner similar to C_3A, both in the presence and absence of gypsum. C_4AF hydrates rapidly (although not as fast as C_3A) to form crystalline materials, which result in the formation of a solid solution $C_3(AF).6H$. The rate of the hydration is seldom sufficiently rapid to cause a flash set, and the effect of gypsum on the rate of reaction is greater than that on C_3A. As the iron content of this phase increases, the rate of hydration decreases. In the reaction products, iron oxide plays the same role as alumina (Fe_2O_3 is substituted for Al_2O_3) (Young, 1981).

3.2 MORPHOLOGY OF HYDRATED CEMENT PRODUCTS

3.2.1 Gels (CSH)

Many **CSH** gels have been identified; the morphology changes with time, and also with the conditions of the mix, for example the amount of water available. The gels can be fibrous or in the form of about 1 nm thick layers which may be rolled up, distorted or stacked up. The gels coat the grain surface. They may be completely amorphous, or consist of very fine crystallites. It is these gels which provide long-term strength to the set cement.

3.2.2 Calcium hydroxide (CH)

Calcium hydroxide is formed as large laths or plates which have been precipitated from solution.

3.2.3 Hydrated C_3A reaction products

In the presence of gypsum, AF_t, the first reaction product of C_3A, forms a coating on the calcium aluminate particles which results in the particles still being able to slide over one another and the mix is still workable.

After a few hours the ettringite crystals form a needle-like structure, and these needles give the initial strength to the cement, but they change to hexagonal plates at a later stage as the monosulphoaluminate phase is formed. In the absence of gypsum, lath-like hexagonal crystals are formed which interlock and result in a flash set; these convert to cubic crystals at a later stage, with an accompanying reduction of volume, increased porosity and potential loss of strength (see Chapter 15 on high alumina cement).

3.2.4 Hydrated C_4AF reaction products

The reaction products in the absence and presence of gypsum have a morphology similar to those of **C_3A**.

3.3 HYDRATION OF CEMENT PASTE

The hydration reactions that occur when Portland cement is mixed with water are obviously complicated because of the multiphase nature of the cement clinker. Only a brief description will be given here; more complete details are given by Scrivener (1989). Once the cement is mixed with water, heat is evolved and the rate of heat evolution can be used as an indication of the reactions that are taking place. Three major steps can be identified.

(a) (0–~3 hr)

During this time a large amount of heat is evolved. The rate of heat evolution is originally high but then rapidly decreases to a minimum after about three hours. The cement remains fluid and workable. The **C_3S** becomes coated with a gelatinous layer which has the morphology of exfoliating films. Small rods of AF_t are observed to form outside this layer. At the end of this period, the **CSH** layers start forming fibrils.

(b) (~3–24 hr)

The rate of heat evolution again increases and about 30% of the hydration of the cement occurs. Rapid growth of both the **CSH** and **CH** hexagonal platelets takes place. Small grains of cement are covered with fibrils to form rosettes or spherulites, and long rods of AF_t develop after about 16 hours. The growth of the **CSH** between the grains results in the setting of the cement paste. At the end of this time, all grains of diameter less than about 5 mm have completely hydrated. Those grains that originally contained some aluminate phase are now hollow shells.

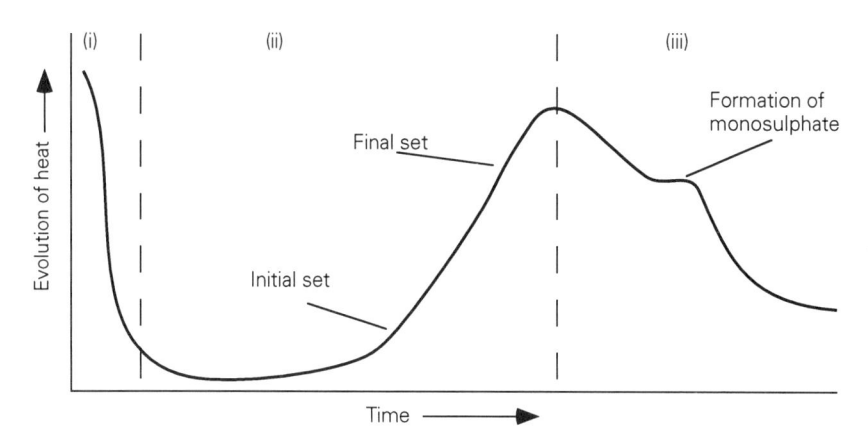

Figure 3.4 Rate of heat evolution from Portland cement

(c) (after 24 hours)

The rate of heat evolution declines, but hydration may continue for years. Further hydration of the alite phase leads to the deposition of **CSH** on the inside of the hollow shells. The belite phase reacts only slowly, and only a small amount of the hydration products are formed even after several months.

A schematic representation of the stages of cement hydration as indicated by the rate of evolution of heat is shown in Figure 3.4.

3.3.1 Rate of hydration

There are many factors that affect the rate of hydration of the various phases in cement, and no two samples of cement will behave in an identical manner. The reactivity of the phases is dependent on such factors as the mean particle size and particle size distribution. These, in turn, are dependent on the degree of grinding of the clinker, its rate of cooling, the impurities present, the relative amounts of the phases and the interaction between the phases as hydration proceeds (Mindess, 1983). For example, the rate of hydration of C_2S is increased in the presence of C_3S due to changes in the concentration of Ca^{2+} and OH^- in the solution from the hydration of the C_3S. The rates of hydration of the other components are also interconnected. However, in general, the rates of reaction will be in the order $C_3A>C_3S>C_4AF>C_2S$. This is indicated by the depth of hydration product formed after six months on 30–55 μm grains of each phase in Table 3.3.

Table 3.3 Depth of hydration product (μm) on 30–55 μm grains (Lea, 1970, p.240)

	3 days	7 days	28 days	6 months
C_3A	10.7	10.4	11.2	15
C_4AF	7.7	8.0	8.4	13.2
C_3S	3.5	4.7	7.9	15.0
C_2S	0.6	0.9	1.0	2.7

Note: The hydration reactions shown above are for the pure compounds. The rapid hydration of C_3A and C_4AF is controlled by the addition of gypsum.

3.3.2 Heats of hydration

All the setting reactions are exothermic, i.e. heat is evolved, but the amount of heat evolved (ΔH) is different for each the reaction, as shown below.

C_3S	–114 kJ/mol
C_2S	–43 kJ/mol
C_3A (with gypsum)	–362 kJ/mol
C_4AF	–203 kJ/mol

These figures represent the final amount of heat evolved for each reaction, but in the hydration of the cement, it is of interest to know the amount of heat being liberated at any given time after hydration has commenced (Mindess, 1983). Expressions have been derived to determine the total amount of heat liberated by each component up to a given time, as shown below:

$$-\Delta H_{(3\ days)} = 240(C_3S) + 50(C_2S) + 880(C_3A) + 290(C_4AF)$$
$$-\Delta H_{(1\ year)} = 490(C_3S) + 225(C_2S) + 1160(C_3A) + 375(C_4AF)$$

where the quantities of the components are expressed in weight fraction, and ΔH in kJ/kg. These equations clearly show the importance of the C_3A, C_4AF and C_3S phases in the amount of heat evolved soon after hydration commences.

3.3.3 Volume changes during hydration

One of the characteristics of Portland cement is that there is very little change in total volume as the cement/water mixture sets. This is because the hydration products are less dense than the unhydrated compounds and fill the volume that was previously occupied by both the water and the unhydrated compounds. It should be noted that the hydration reactions continue long after the cement has set and it is here that the difference in the hydration behaviour of the compounds becomes important.

The grains of the major components, **C₃S** and **C₂S**, will continue to hydrate after the cement has set, but the hydration products will only occupy the space available for them, that is, the volume occupied by the water with which they are reacting. As soon as this volume is filled with reaction products, the reaction ceases.

However the hydration products of the minor components, **C₃A** and **C₄AF**, continue to be formed even when there is insufficient space and the hydration of these compounds results in an expansion of the paste volume. If the paste has already hardened, then the hydration of these compounds results in a build-up of stress in the set cement which will cause cracking. This is why the addition of the correct amount of gypsum to the cement paste is critical. If too little is added then a flash set will occur. If the correct amount is added, then the expansion caused by the hydration of **C₃A** will take place before the cement has set, and no damage is done, but if too much gypsum has been added, then ettringite will continue to be formed after the cement has set and cracks are then formed.

It should be emphasized here that **C₃A** and **C₄AF** are minor constituents in the cement and therefore the net expansion that takes place when they hydrate is small, and, as will be seen later, this can be put to advantage to counteract the drying shrinkage of the cement.

The amount of water theoretically required to hydrate the cement paste completely is about 0.25 times the mass of the cement. If this amount is used it is found that there is still some unreacted cement. This is because some of the water becomes adsorbed in the pores between the gel particles (gel pore) and is therefore not available for reaction with the cement. In reality, in order to get complete hydration a water/cement ratio of about 0.38 has to be used in the mix.

3.3.4 Summary of the hydration of Portland cement

- It is the formation of **CSH** which gives the cement its strength. The laths and needles formed due to the reactions of **C₃A** and **C₄AF** result in the initial set of the cement, but these do not contribute significantly to the long-term strength.
- **C₃S** forms **CSH** much more rapidly than **C₂S**; therefore cements with a higher proportion of **C₃S** will develop strength more rapidly than cements with less **C₃S**.
- While it is agreed that the formation **CSH** gels are the major factors which contribute to the strength of cement, the role of the **CH** platelets is still uncertain.

- The rates of hydration are important; C_3A, C_4AF and C_3S all hydrate rapidly, but C_3S is responsible for the major part of the early strength developed. C_2S hydrates more slowly, and it is this that gives the cement its increase in strength over the long term.
- It is the hydration of C_3S and C_3A that is responsible for most of the heat evolved in the cement in the first 48 hours. Therefore if this heat evolution is to be limited, C_3A should be replaced by C_4AF, and the amount of C_3S should be reduced (this involves a corresponding increase in C_2S, which results in a reduction in early strength but not in the ultimate strength of the cement).

3.4 TYPES OF PORTLAND CEMENT

3.4.1 Classification of cements

By adjusting the relative amounts of the phases present in Portland cement, the properties of the cements can be altered. Different types of cement are produced. The classifications shown below are those used in Australia (Table 3.4 (a)) and America (Table 3.4 (b)). (However, the

Table 3.4 (a) Types of Portland cement produced in Australia (Cavanagh and Guirguis, 1992)

Type	C_3S	C_2S	C_3A	C_4AF	Common designation
A	48–65	10–30	2–11	7–17	'ordinary'
B	50–65	7–25	6–13	7–13	'rapid set'
C	25–30	40–45	3–6	12–17	'low heat'
D	50–60	15–25	2–5	10–15	'sulphate resist'

Hypothetical phase composition (mass %) (range)

Table 3.4 (b) Types of Portland cement produced in USA (Mindess, 1983)

Type	C_3S	C_2S	C_3A	C_4AF	$C\bar{S}H_2$ added	Heat evolved (7 day, kJ/kg)	Common designation
I	50	25	12	8	5	330	'ordinary'
II*	45	30	7	12	5	250	
III	60	15	10	8	5	500	'rapid set'
IV	25	50	5	12	4	210	'low heat'
V	40	40	4	10	4	250	'sulphate resist'

Hypothetical phase composition (mass %)

*Type II has moderate sulphate resistance and low heat combined with normal gain in strength.

TYPES OF PORTLAND CEMENT

Australian classifications have recently been altered. The new classifications are described in Appendix A.)

It is important to realize that the relative amounts of the phases present are calculated using the Bogue equations (see page 29) from chemical analysis data and the assumption is made that chemical equilibrium has been reached in the formation of the cement clinker. This takes no account of the formation of non-equilibrium reaction products such as glass and also the formation of solid solutions in the Portland cement clinker phases. The Australian standard AS 2350.2–1991 has the disclaimer that 'The hypothetical compound composition percentage (by mass) calculated – from the chemical analysis does not imply that the oxides are actually or entirely present as such compounds or that such compounds are present in the percentages calculated.'

There is also an extremely wide range in the amounts of each phase present in cements of each type, as is shown in the Australian data in Table 3.4 (a). This is mainly due to variations between different manufacturers who use different raw materials and clinkering processes. An example of the variations in some British and American Portland cements obtained from different manufacturers is given by Lea (1970, pp. 159–60). The standards only put limits on the amounts of phases that are critical in obtaining the desired properties.

3.4.2 Properties and uses of the types of Portland cement

(a) Type A – Ordinary Portland cement (USA type I)

This is the most common type of Portland cement that is used for general construction purposes where there is no exposure to sulphates in the soil or in the ground water.

(b) Type B – Rapid-set cement (USA type III)

There is not much difference in the chemical composition between this and ordinary Portland cement. The rapid-set properties are due mainly to the greater fineness of the cement powder and also, to a lesser extent, the higher **C_3S** content. The principal reason for its use is that the rapid-setting properties mean that formwork can be removed early for re-use, and it is also useful in cases where sufficient strength for further construction is required quickly. It is used for sea walls, piers, thin panels, etc.

(c) Type C – Low heat cement (USA type IV)

This was developed in the USA for use in the construction of massive dams. Because of the lower **C_3S** and **C_3A** content, there is a slower development of strength than with ordinary cement, but the ultimate strength is the same. It is used for large constructions or the placement of concrete in very hot weather, where heat evolved during setting and curing is difficult to remove and the temperature rise in the concrete can become excessively large.

(d) Type D – Sulphate resist cement (USA type V)

Sulphates, present in ground water, can attack cement. In general the reaction between the sulphates and the set cement forms products which cause cracking of the set cement. The extent of the attack is dependent on the type of sulphate present in the water (calcium, sodium or magnesium).

Note: **N** = Na_2O and **M** = MgO in the abbreviated formulae below.

(i) $CaSO_4.2H_2O$ ($C\bar{S}H_2$, gypsum)
Sulphate attack occurs by the reaction of gypsum with tricalcium aluminate.

$$C_3A + 3C\bar{S}H_2 + 26H \rightarrow C_6A\bar{S}H_{32} \qquad \text{ettringite}$$

This is the same reaction that was used to prevent the 'flash set'. The attack is limited due to the formation of the insoluble calciumaluminosulphate around the calcium aluminate grains.

(ii) $Na_2SO_4.10H_2O$ ($N\bar{S}H_{10}$, Glauber's salt)
Firstly sodium sulphate reacts with the calcium hydroxide (**CH**) platelets which form during the hydration of **C_3S**.

$$CH + N\bar{S}H_{10} \rightarrow C\bar{S}H_2 + NH + 8H$$

Secondly the calcium sulphate reacts with the **C_3A** to form ettringite. The water will eventually become saturated with sodium hydroxide and the reaction will cease. The reaction will only proceed if the soluble sodium hydroxide (**NH**) is removed from the water as it is produced. The attack is also limited by the formation of the insoluble calciumaluminosulphate.

(iii) $MgSO_4.7H_2O$ ($M\bar{S}H_7$, Epsom Salts)

$$CH + M\bar{S}H_7 \rightarrow C\bar{S}H_2 + MH + 5H$$

Magnesium sulphate reacts in a similar way to sodium sulphate, but in this case, the magnesium hydroxide (**MH**) that is produced has a very low solubility in the water, and therefore the equilibrium is forced to the right, and the reaction proceeds. The situation is worsened by the fact that the calciumaluminosulphate is unstable in the presence of magnesium sulphate, and by the continued action of this salt is decomposed again to form gypsum and hydrated alumina, thus exposing fresh surface for the attack to continue.

It has been found that a reduced C_3A content does lead to a better resistance of the cement to attack by magnesium sulphate.

QUESTIONS

1. Explain the differences between the following (as applied to Portland cement):
   (a) hardening and setting,
   (b) false set and flash set, and
   (c) Type A and Type C cement.
2. What effects do you think the particle size of the cement would have on the setting reactions?
3. What hydration reactions are mainly responsible for the setting of cement?
4. Is there any relationship between the heat of hydration and the cementing properties of the components in cement?
5. Why is the presence of C_3A undesirable in cement? What is done to help counteract these problems?
6. Calculate the amount of heat evolved up to three days and up to one year after hydration for the four cement compositions given in Table 3.5 (below). Comment on the significance of these results.
7. Gypsum is added to cement to prevent a flash set, and yet it is known that it can cause sulphate attack on cement. Explain this apparent contradiction.
8. Why is it difficult to use electron microscopy (SEM and TEM) to determine the morphology of the hydration products of cement?

Table 3.5 Compositions of four cements (wt%)

Cement	C_3S	C_2S	C_3A	C_4AF
1.	50	25	12	8
2.	60	15	10	8
3.	25	50	5	12
4.	40	40	4	10

References

Cavanagh, K. J. and Guirguis, S. (1992) Cements. In W. G. Ryan and A. Samarin (eds), *Australian Concrete Technology*, Longman-Cheshire, Melbourne, pp. 1–13.

Jawed, I., Skalny, J. and Young, J. F. (1983) Hydration of Portland cement. In P. Barnes (ed.), *Structure and Performance of Cements*, Applied Science, London, pp. 237–317.

Kohlhaas, B. (1983) *Cement Engineers' Handbook*, 4th English edn, Bauverlag, Wiesbaden, pp. 128–32.

Lea, F. M. (1970) *The Chemistry of Cement and Concrete*, 3rd edn, Arnold, London.

Mindess, S. (1983) Concrete materials, *Journal of Materials Education*, **4**, 984–1046.

Scrivener, K. L. (1989) The microstructure of concrete. In J. P. Skalny (ed.), *Materials Science of Concrete I*, The American Ceramic Society, Westerville, OH, pp. 127–61.

Young, J. F. (1981) Hydration of Portland cement. *Journal of Education Modules for Materials Science and Engineering*, **3**, 403–28.

Mortar 4

Mortars are used to bed and joint building units (e.g. bricks) to give structural strength and to exclude rainwater. The mortars used until 70 or 80 years ago were sand-lime mortars. These have now been replaced by cement-plasticizer-sand or cement-lime-sand mixtures. The relatively slow-paced construction methods used at the start of the twentieth century meant that the sand-lime mortars, which develop strength slowly, could be used. These mortars were made from slaked lime ($Ca(OH)_2$) ground to a very fine powder which was then mixed with sand and water at the building site. The slaked lime forms a paste of colloidal and crystalline calcium hydroxide. Evaporation of water results in the setting of the paste, but hardening only occurs slowly by the reaction of the calcium hydroxide with carbon dioxide from air. Such a mortar will not harden in a wet environment (Sahlin, 1971). The introduction of Portland cement into the mortars resulted in a mortar that would develop strength more rapidly and that would set in water. This meant that more rapid construction was possible. Sand-lime mortars are now rarely used (Lenczner, 1972).

The proportioning of the components in the mortar is usually by volume, since this can be done on site with the simplest of equipment. The proportion of lime or cement to sand is normally 1:3 by volume. The properties of the mortar can be altered by changing the relative amounts of lime and cement in the mortar.

4.1 PROPERTIES OF WET MORTAR

The ideal mortar possesses good 'workability' while it is wet, and when set attains the strength and bonding characteristics required for structural stability and durability. The workability of a mortar must be such that it can fill all the joints easily and when a course of units has been laid, the mortar should be rigid enough to support the next course, but still be deformable so that any necessary corrections can be made to the newly laid units. Workability is almost impossible to define precisely and to measure. The workability of a mortar is dependent on its spreading characteristics, its ability to cling to a vertical surface and its resistance to flow after being placed in between bricks. The addition of lime to a Portland cement mortar improves the workability of the mortar (Lenczner, 1972).

The water retention of wet mortar is important. If the water retention of the mortar is low and if it is placed on a water absorptive unit (e.g. an underfired porous clay brick) the mortar will lose water rapidly and become very difficult to work. It is also possible that the mortar will be dried out to such an extent by the lower unit that the new unit will be laid in a partly dried and set mortar bed. This will result in a low-strength mortar joint and cracks will soon appear. Even if the water retentivity is improved, the thin layer of mortar adjacent to the porous brick might dry too quickly and cracks will appear between the units and the hardened mortar joints. Water retention can be improved by the addition of air-entraining agents and finely ground inorganic plasticizers such as limestone, clay or lime, or organic materials such as saponified pitch, naphthanate soap etc. (Dmitriev and Kotov, 1971). The use of polymer additions to mortar to improve workability and water retention will be discussed in Chapter 15.

4.2 STRENGTH OF MORTAR

The setting and the strength of the cement-based mortars is due to the hydration of the cement paste. The strength will be affected by the cement content of the mix, the water/cement ratio, the proportion of cement to sand and the properties of the sand. High compressive strength is generally achieved by using a high cement content, a low water/cement ratio, and a coarse sand.

Natural sand is normally used in mortars, but its particle size range and distribution should be controlled. The particle size range is normally between 75 μm and 2 mm. Spherical or rounded sand particles produce a mortar which is easy to work. The maximum particle size of the sand should be no more than one-third to one-half of the thickness of the joint.

The hardened mortar in a completed structure must be able to transfer the compressive, tensile and shear stresses between adjacent units, but the weakest mortar that can fulfil these functions should be used. The reason for this is that if there is any movement due to thermal expansion or contraction, settlement, shrinkage or other causes, it is desirable for the cracks to appear in the mortar rather than in the units. These cracks tend to be smaller and much easier to repair than cracks in the bricks (Curtin et al., 1982).

Table 4.1 Typical mortar mixtures (Curtin *et al.*, 1982)

Mortar group	Cement: Lime: Sand	Masonry cement: Sand	Cement: Sand with plasticizer
I	1 : 0–0.25 : 3		
II	1 : 0.5 : 4–4.5	1 : 2.5–3.5	1 : 3–4
III	1 : 1 : 5–6	1 : 4–5	1 : 5–6
IV	1 : 2 : 8–9	1 : 5.5–6.5	1 : 7–8
V	1 : 3 : 10–12	1 : 6.5–7	1 : 8

(Masonry cement = Portland cement + lime in equal proportions.)

4.3 MORTAR MIXES

Typical mortar mixes (proportions by volume) are shown in Table 4.1.

The strength of the mortars decreases with increasing lime and sand content (from group I to V) and thus the ability to accommodate movements due to settlement and shrinkage increases. The strength within each group of the mixes is the same, but the frost resistance increases from left to right whereas the improvement of the bond and resistance to rain penetration increases from right to left. The factors affecting frost resistance are discussed in more detail in Chapter 12, but, in general, an increase in porosity due to entrained air will increase the resistance to failure by successive freeze/thaw cycles of the water in the pores of the mortar.

Type I mortar group is too strong for general use as its compressive strength is similar to that of medium-strength fired clay bricks. The strength cannot be decreased by the addition of more sand, since this would result in the production of an unworkable mix because there would be insufficient fine cement in the mixture. Therefore the desired reduction in strength is achieved by the partial substitution of cement by fine slaked lime.

The compressive strength of a pure cement mortar is about 20 times that of a pure lime mortar. In recent times plasticizers other than lime have been used with Portland cement. These include clay, limestone and air-entraining agents, and result in improved workability characteristics without affecting the hardening or setting properties. Glue is sometimes added to the lime-cement-sand mortar to make a rapidly setting and strong mortar, but the costs are high, and this mortar is only used for thin joints.

4.4 SPRAYED MORTAR

The spraying of mortar mixes onto a surface by compressed air at high velocity is a well-established technique (e.g. sprayed onto a rock face in a tunnel or as a method of repair). The sprayed material is variously known as 'Gunite concrete' or 'Shotcrete'. No formwork is required and in some cases no reinforcement is necessary. There are two major methods of applying Shotcrete, namely dry-mix or wet-mix. The dry mix process involves the premixing of the dry ingredients which are then blown through a hose by compressed air into a nozzle where water is added to the mix just prior to spraying, whereas in the wet-mix process, the water is added to the premixed components in a mixer and then the slurry is sprayed through a nozzle with compressed air. The difference in mixing methods results in the wet mix always requiring a higher water/cement ratio than the dry mix, even though this can be lowered by the use of high-range water-reducing admixtures. This results in the wet mix being more prone to shrinkage and having a higher porosity and permeability than the dry mix. Utilization of the dry-mix method means that only sufficient water need be added for the hydration of the cement and the mix has very little shrinkage, a very low permeability and consequently high durability (Warner, 1995).

QUESTIONS

1. Describe the setting reactions of sand-lime and cement-sand-lime mortars.
2. What is the effect of the addition of lime to a cement-sand mortar?
3. Why is the water retention of a mortar important and how can it be improved?
4. What are the major effects of increasing the lime and sand content of a cement based mortar?
5. Why is sand added to mortars and what effect does the shape of the sand particles have on the mortar?
6. What are the required properties of a mortar, and why should the weakest mortar that meets these requirements be selected for use?

References

Curtin, W. G. *et al.* (1982) *Structural Masonry Designers' Manual*, Granada, London, pp. 452–55.

Dmitriev, A. S. and Kotov, I. T. (1971) Materials for masonry structures. In S. A. Sementsov, and V. A. Kameiko (eds), *Designer's Manual, Including Reinforced Masonry*, Keter Press, Jerusalem.

REFERENCES

Lenczner, D. (1972) *Elements of Load Bearing Brickwork*, Pergamon Press, Oxford, Ch 2.

Sahlin, S. (1971) *Structural Masonry*, Prentice-Hall, Englewood Cliffs, Ch B.

Warner, J. (1995) Understanding shotcrete – the fundamentals. *Concrete International*, **17** (5), 59–64.

5 Concrete

Concrete is a combination of mortar (cement, fine sand aggregate and water) and coarse aggregate.

The compressive strength of set concrete is dependent mainly on the type of the cement in the mortar, the type of aggregate, the cement/aggregate bond, the water/cement ratio used in the concrete mix and the degree of compaction of the wet concrete.

As well as the factors which affect the compressive strength of the set concrete, it is also necessary to investigate the factors that affect the ease of pouring of the concrete, i.e. the workability of the concrete.

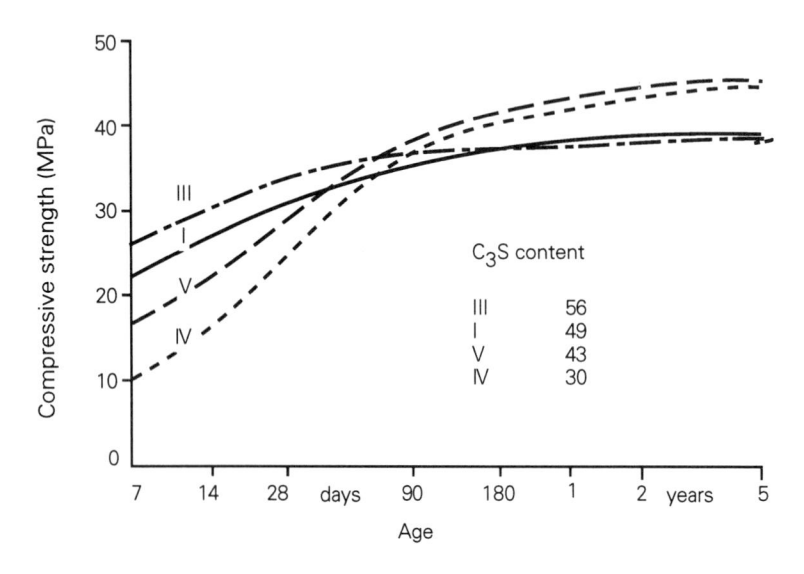

Figure 5.1 Strength development of concretes made with cements of different types (redrawn from Neville, 1981, p. 65)

AGGREGATE AND THE AGGREGATE/CEMENT BOND

5.1 EFFECT OF TYPE OF CEMENT ON STRENGTH

The factors affecting the strength of mortars have already been mentioned in Chapter 4 and have been seen to include the type of cement that is used, the water/cement ratio, the cement/sand ratio and the physical properties of the sand.

Portland cement is normally used to make ordinary strength concrete and the development of strength over time is dependent on the type of Portland cement used. (The types of Portland cement used in Australia and the USA are listed in Tables 3.4 a and b and in Appendix A.)

The relationship between the concrete strength and the types of cement used to make concrete is shown in Figure 5.1. The ASTM system of naming Portland cements is used. (It should be noted that the rate of development of strength up to 90 days increases with the **C_3S** content of the cement.)

5.2 AGGREGATE AND THE AGGREGATE/CEMENT BOND

Aggregate is much cheaper than cement, therefore as much as possible is put into the concrete mix. At least 75% of the volume of the concrete is composed of aggregate, so the aggregate can be regarded as a building material which is bonded together by the cement. This means that aggregates are not simply 'inert fillers'; their physical, thermal and often also their chemical properties profoundly influence the durability and strength of concrete. The aggregate materials usually have a higher volume stability and better durability than the cement alone.

The cement serves to bond the aggregate particles together and the nature of the interface between the aggregate and the cement is of great importance. The interfacial region between the aggregate and the cement is thought to be very different from the bulk cement regions both in terms of morphology, density and composition (Mindess,1989). These regions are normally lower in density than the hydrated cement matrix and contain large orientated hexagonal crystals of **CH** and needles of AF_t. The structure is shown schematically in Figure 5.2 (redrawn from Mindess, 1989).

The thickness of this interfacial region is about 50 μm, with the weakest part of the zone lying about 5 to 10 μm away from the interface. The existence of this porous region can effect the fracture characteristics of the concrete and be a factor that limits the strength of the concrete.

Some idea of the quality of the bond between the aggregate and cement can be obtained by inspecting a fractured sample of concrete. Ideally, there should be some fractured aggregate and some pull out. If all the aggregate has fractured, then the aggregate is too weak, and if all the aggregate has pulled out, then either the bond between the cement paste (mortar) and the aggregate is too weak or the cement paste itself is too weak.

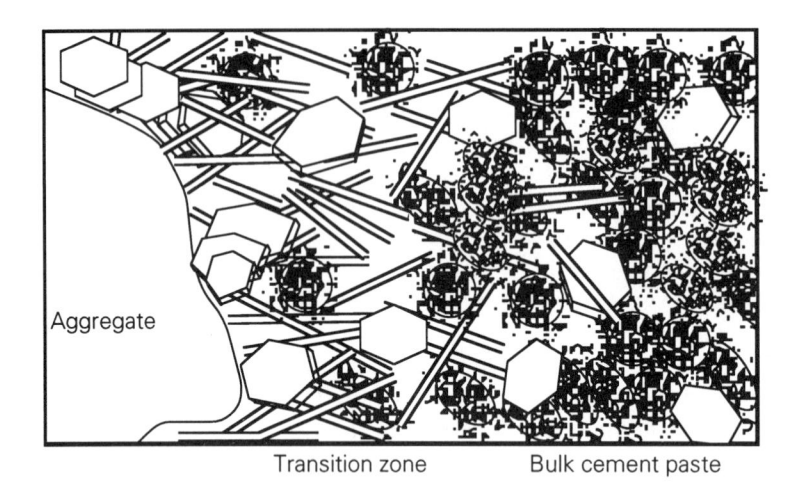

Figure 5.2 The interface between the aggregate and cement paste

5.2.1 Aggregate materials

Aggregates are normally divided into two classes, namely dense, which is used for the production of structural high load-bearing and reinforced concretes, and lightweight, which is used in situations where sound and thermal insulation are major requirements.

The density of ordinary concrete, made with dense aggregate, is 2240–2400 kg/m^3, and that of lightweight concrete, made with lightweight aggregate, is 640–1600 kg/m^3.

(a) Dense aggregate materials

In general, these materials must be non-porous and chemically stable in the cement environment. It is also advantageous if the cost of the material is low, hence the use of natural rocks and minerals, or waste materials. The four major classifications of dense aggregates are sands and gravels, rocks, blast furnace slag and broken bricks.

Sands and gravel consist of minerals that are resistant to weathering over long periods. Sands (with particle diameter < 5 mm) are normally quartz (SiO_2), whereas gravels (with particle diameter >5 mm) may consist of quartz, quartzite, granite, sandstone, limestone etc. Some of the various types of crushed rock that can be used as aggregates are listed in Table 5.1.

Air-cooled blast furnace slag, which is obtained as a waste product from the manufacture of pig iron, can also be used as an aggregate but care

Table 5.1 Types of rock for use as dense aggregate material

Rock	Comments
Granites	Well crystallized, fine to coarse grain
Dolerites	Very fine grain crystallized
Basalts	Very fine grain cystallized
Sandstones	Cemented quartz grains
Limestone ($CaCO_3$)	Must be dense and hard
Dolomite ($CaCO_3.MgCO_3$)	Must be dense and hard

must be taken in the preparation and selection of the slag. It must be crystalline (not glassy) and chemically stable. Dense blast furnace slag is produced by very slow cooling of the molten slag in air. The chemical composition of the slag is important; it must not contain constituents which would cause problems with the set concrete. Examples of problem constituents are dicalcium silicate, sulphur compounds, high ferrous iron content or alkalies. Dicalcium silicate (**C_2S**), can cause 'dusting' or 'falling' unsoundness in the aggregate because it undergoes phase changes at it is cooled. α-**C_2S** transforms to α'-**C_2S** at 1400 °C which then transforms to either the stable γ-**C_2S** at 850 °C or the metastable β-**C_2S** below 675 °C. If the metastable β-**C_2S** is formed then it can revert to the γ-**C_2S** at ambient temperatures at some later time. This transformation is accompanied by a 12.5% increase in volume and can cause the aggregate to crumble in the set concrete. The other problem constituents react chemically with the set concrete, and can result in sulphate or alkali attack.

Broken bricks can be used as dense aggregate material, provided the calcium sulphate (gypsum) content is low (this can be present if the bricks come from demolition sites and are mixed up with gypsum plaster). The presence of gypsum would cause the concrete to expand and crack due to sulphate attack after the concrete has set.

The types of materials that may be used for aggregates are fully specified in standards, e.g. ASTM C33-82, AS 2758.1.

(b) Lightweight aggregates

Natural and artificial porous materials are used to make lightweight concrete. These include pumice; clinker; expanded clay, shale and slate; exfoliated vermiculite and expanded volcanic glass (perlite). Foamed blast furnace slag can also be used as a lightweight aggregate. This is formed by foaming the molten slag with water sprays to generate steam, and then cooling slowly to ensure the slag has time to devitrify (crystallize).

A more detailed description of these porous materials is given in standards (e.g. ASTM C 330-89 for lightweight aggregate in structural concrete).

Table 5.2 Maximum size of aggregate and size of article

Minimum dimension of section	Maximum size of aggregate
6–12 mm	2 mm
12–26 mm	4 mm
26–70 mm	6 mm
>70 mm	112 mm

5.2.2 Aggregate particle characteristics

(a) Aggregate size

In a concrete mix, the maximum size of the aggregate particles used is dependent on the size of the article being made, as shown in Table 5.2.

In general, when considering reinforced concrete the aggregate particles should not be larger than three-quarters of the space between reinforcing bars and one-fifth to one-third of the minimum wall thickness. If larger aggregate is used then the continuous zone of weakness adjacent to the aggregate surface would then occupy a large fraction of the wall thickness.

(b) Aggregate shape, texture, porosity and the grading of the aggregate

Ideally, in mixed concrete, each piece of aggregate (large or small) should be totally surrounded by cement, and the packing of the particles should be as dense as possible. The grading of the aggregate (the selection of the particle size ranges to be used and the amount of aggregates in each range) should be such as to give the maximum density of particles packing which minimizes the amount of the more expensive cement used in the mix.

Aggregate shape
The relative quantities of coarse and fine aggregate needed to give dense packing will depend on the shape of the aggregate particles. If the aggregate is angular, then the relative proportions of coarse/fine will be different than if it is rounded (spherical), because the random packing of coarse non-spherical particles results in a greater volume of voids than the packing of coarse spherical particles and the voids need to be filled by fine particles.

In general, angular aggregates will produce concrete of a greater strength than spherical aggregates. This is due to the greater surface area/volume ratio of the angular aggregates which produces a larger bonding interface between the aggregate and the cement.

Surface texture
The surface texture of the aggregate is also important, a rough texture results in better bonding because of the larger surface area of the aggregate in contact with the cement.

Porosity
The absorption of water by the aggregate must be taken into account when designing a concrete mix, because if dry porous aggregate is used it will absorb water during the mixing of the concrete and lead to a reduction in the amount of water available to hydrate the cement.

Grading of the aggregate
The aim in the grading of the aggregate is to minimize the void space between the aggregate, since this has to be filled with the cement.

The selection of the relative amounts of the coarse and fine grades of aggregate is dependent on many factors, which include the maximum size of the aggregate, the shape of the aggregate, the size distribution of the coarse and fine fractions, the workability required, the surface texture of the aggregate, the water/cement ratio to be used and the required strength of the concrete.

Various codes exist that deal with the aggregate grading and empirical and calculated tables and curves have been constructed to account for all the variables. Examples of such data are given by Neville (1981, Ch. 10).

It is found that if the aggregate is graded to give the maximum density of packing then an unworkable mix is produced. Workability is generally defined as the ease with which the concrete can be mixed, handled, transported, placed and compacted with a minimum loss of homogeneity. For a concrete to be workable, it must contain an amount of material of particle size < 300 μm in excess of that required to completely fill the voids between the coarse aggregate. This material consists of fine aggregate (sand) and cement and acts as a 'lubricant' between the larger particles. This excess should be kept to a minimum to minimize the overall cost of the concrete.

5.2.3 Major harmful impurities in the aggregate

Impurities may interfere with the hydration of the cement, may coat the aggregate preventing the formation of a good bond between the cement and the aggregate, or may react chemically with the cement.

The hydration of the cement can be affected by organic impurities which are normally present as products of decayed vegetable matter. These may produce tannic or humic acids which can interfere with hardening and strength development in the concrete. Organic impurities are most likely to be found in the fine aggregate (sand).

Clays, if present as 'lumps', may break down after the concrete has set. This results in the formation of pits in the concrete and may also give a poor surface. If clays are present in the aggregate 'fines' then the clay may form a coating on other aggregate particles thus preventing the formation of a good aggregate/cement bond.

Materials that can react chemically with the cement include some oxides (CaO, FeO, MgO). These cause problems since they can react with water to form the hydroxides, which cause expansion after the concrete has hardened and result in cracking. CaO has been found in blast furnace slags and expanded shales.

Silica (SiO_2) is dangerous if present in its 'active' form, e.g. as opal, chalcedony (cryptocrystalline fibres) and tridymite. In these forms a reaction can take place between the silica and alkali impurities in the cement producing an alkali-silica gel. This gel absorbs large amounts of water and swells within on the set cement. The swelling results in the cracking of the concrete from the alkali/aggregate reaction. Glassy materials react in a similar manner to the active forms of silica and form alkali-silicate gels with sodium or potassium hydroxides

Iron sulphide, FeS_2 (marcasite or pyrites), can react with alkalis to form $Fe(OH)_2$:

$$FeS_2 + O_2 + 2Ca(OH)_2 + H_2O \rightarrow Fe(OH)_2 + 2CaSO_4 \cdot 2H_2O$$

The calcium sulphate can then react with the set cement and cause expansion and cracking by sulphate attack.

5.3 EFFECTS OF WATER/CEMENT RATIO AND WORKABILITY ON STRENGTH

For a given type of cement and given cement/aggregate proportions the compressive strength of 'normal strength' set concrete at a given age is primarily dependent on two factors, namely the water/cement ratio used in the mix and the degree of compaction of the wet concrete.

5.3.1 Water/cement ratio

The effect of the water/cement ratio on the strength of concrete can be described by two 'laws' (Neville, 1981, p. 268).

Abrams, in 1919, found that the compressive strength, S, of set concrete was inversely proportional to the water/cement ratio used in the concrete mix. Abrams' Law can be expressed as:

$$S = \frac{K_1}{(K_2)^{w/c}}$$

EFFECTS OF WATER/CEMENT RATIO AND WORKABILITY

where K_1 is an empirical constant with the dimensions of S and K_2 is a dimensionless empirical constant and both are dependent on the materials and the test conditions, w/c is the water/cement ratio measured by volume.

This law assumes that the concrete is fully compacted, by which is meant that all of the air that was present between the dry particles of sand, cement and aggregate is removed after the addition of water to the mix. However, in practice, fully compacted concrete will still contain approximately 1% of air voids. Abrams' Law is thus a special case of 'Feret's Law', formulated in 1896 by Feret, which takes the volume of entrained air, a, into account, and can be stated as:

$$S = K\,[c/\,(c + w + a)]^2$$

where c, w and a are the absolute volumes of cement, water and air respectively, and K is an empirical constant with the dimensions of S (Neville, 1981, p. 269).

The water/cement ratio determines the porosity of the hardened cement paste at any stage of hydration, and the degree of compaction will determine the amount of entrained air in the mix. Thus the water/cement ratio and the degree of compaction both affect the volume of the voids in concrete, and this is why the volume of air in concrete is included in Feret's law.

As stated in Chapter 3, the theoretical water/cement ratio (by weight) required for the complete hydration of the cement in the concrete is approximately 0.25, but if this ratio is used there will be some unreacted cement remaining in the paste because some of the water will be trapped in the gel pores in the hydrated cement. Such a mix is also unworkable, compaction is very difficult and this results in air entrainment in the concrete mix. In practice, if allowance is made for bleeding (separation of some water to the cement surface due to settlement of aggregate/paste particles), and for the water trapped in the gel pores in the hydration products, water/cement ratios > 0.38 are required for the complete hydration of the cement. The hydration products of the major components of the cement can only form in the volume occupied by the water and as the cement hydrates, the volume of the gel occupies about double the volume of the dry cement.

If the water/cement ratio is > 0.38, then the volume of the gel formed is insufficient to fill the space available to it, and some unreacted water remains in excess of that required for hydration. The volume occupied by the excess water results in the formation of capillary pores in the set cement. It is the production of these capillary pores which result in the reduction of the compressive strength of the compacted concrete as the water/cement ratio is increased above 0.38. It is found that 2% voids can result in a 10% reduction of strength and 5% voids can result in a

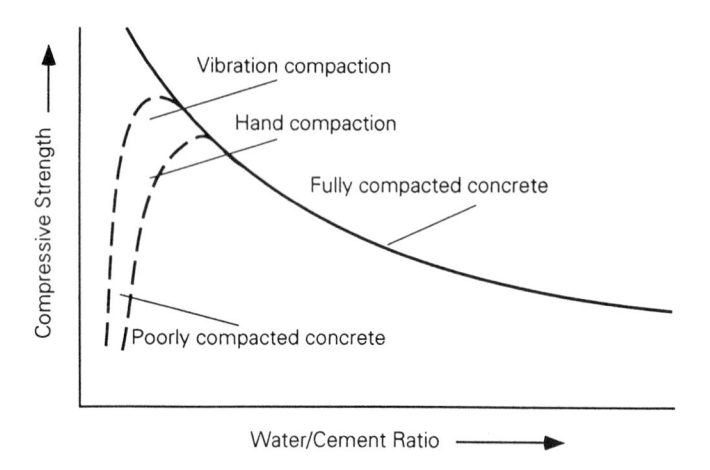

Figure 5.3 Effects of methods of compaction on the compressive strength of concrete for different water/cement ratios (redrawn from Neville, 1981, p. 269)

30% reduction of strength. The effect of such flaws on the strength of concrete is characteristic of materials that exhibit brittle fracture.

At low water/cement ratios, the method used to compact the concrete becomes important, since it determines the amount of air left in the mix. This 'accidental' entrained air was present in the original dry mix as voids within the originally loose granular material. The size of these voids will be governed by the grading of the finest particles in the mix. The voids will be more easily removed by compaction from a wet mix than a dry mix, but their removal is also dependent on the method used to compact the concrete as shown in Figure 5.3.

The reduction in strength of the concrete that occurs at low water/cement ratios is due to the presence of entrained air in the mix while that at high water/cement ratios is due to the presence of capillary pores that were filled with excess water. At low water/cement ratios, hand compaction is less able to remove the entrained air than vibration compaction. However, for either method of compaction, there exists an optimum water content of the mix at which the sum of the volumes of the air bubbles and the water space will be at a minimum, and the strength will be maximized.

Note: The strength vs. water/cement curve assumes that all the water added is available to hydrate the cement. Therefore it is essential that the cement should not dry out while the cement is reacting with the water. Evaporation of the water must be prevented during the curing (setting) of the concrete. It is also assumed that the aggregate particles do not absorb a significant amount of the water added. If the pores in the aggregate are

EFFECTS OF WATER/CEMENT RATIO AND WORKABILITY

not saturated with water prior to mixing, then the effective amount of water available for the hydration of the cement will again be reduced.

5.3.2 Workability

Workability may be described as that property of the plastic concrete mixture which defines the ease with which it can be placed and the degree to which it resists segregation. The ASTM definition is 'that property determining the effort required to manipulate a freshly mixed quantity of concrete with a minimum loss of homogeneity'.

In actual practice the workability required is dependent on the type of construction for which the concrete is to be used and the methods of placing, mixing and working that are to be employed. For example, concrete that can be placed without segregation in a massive dam would be unworkable to form a thin structural member, and concrete that is workable if compacted by high frequency vibrators would be unworkable if hand tamping and spreading was used.

(a) Factors affecting workability

The workability is dependent on many factors which include the proportion of the cement in the mix, the amount of water added, the nature of the sand and aggregate and their relative proportions.

Amount of cement
Very lean mixtures (those made with relatively small amounts of cement to aggregate) tend to produce harsh concrete which has poor workability. Rich mixtures (made with relatively large amount of cement to aggregate) are more workable than the lean, but the concrete containing a very high proportion of cement may be sticky and difficult to finish. The cost of the mix would also be increased.

Consistency
The consistency generally denotes the wetness of the concrete. If the mixture is too wet, then segregation of the aggregate and sand may occur, and conversely, if the mix is too dry, then it is difficult to place. Generally the concrete should have the driest consistency that is practicable for its placement (which is dependent on the method used for placement and compaction). The optimum consistency is determined by the type of structure, type and size of aggregate and the method of compaction to be used.

Sand
The shape and size of the sand particles affect workability. Rounded sand particles gave greater workability than angular sand grains. Concrete

containing fine sand requires more water for the same consistency than an equivalent amount of coarse sand, because of the larger surface area/unit weight of the fine sand. The sand surface has to be wet by the water which acts as a lubricant between the sand particles.

Coarse aggregate

The workability is also affected by the shape and size of the coarse aggregate used in the concrete. Angular, crushed aggregate generally requires more sand and water in the mix to produce the desired workability than concrete made out of rounded aggregate. The increase in water means that if the water/cement ratio is to be kept constant, more cement is required.

As the relative proportion of the mortar (sand, cement, water) is increased, the effects of the grading and angularity of the coarse aggregate becomes less important, but this increases the cost of the concrete.

(b) Measurement of workability

Slump test (Nagarajan and Antill, 1978; Neville and Brooks, 1990)

The slump test is used for measuring the consistency or wetness of concrete. It does not measure all the factors contributing to workability, but it is simple to carry out and is often used as a control test to check for uniformity between concrete batches. The slump test is extremely sensitive to changes in water content, it changes with the tenth power of the change in water content (Popovics, 1994). The test is described in detail in the standards such as ASTM C143-78, or AS 1012.3.

In brief, a hollow truncated metal cone (200 mm bottom diameter, 100 mm top diameter and of height 300 mm) is filled with freshly mixed concrete in three layers. Each layer is tamped 25 times with a standard 16 mm diameter steel rod, rounded at the end. The surplus concrete is removed by means of a screeding and rolling motion of the tamping rod and the cone is then slowly lifted and the unsupported concrete left to slump. The slump is then measured to the nearest 5 mm as the decrease in the height at the centre of the slumped concrete. If, instead of slumping, one half of the cone shears (as shown in Figure 5.4), the test should be

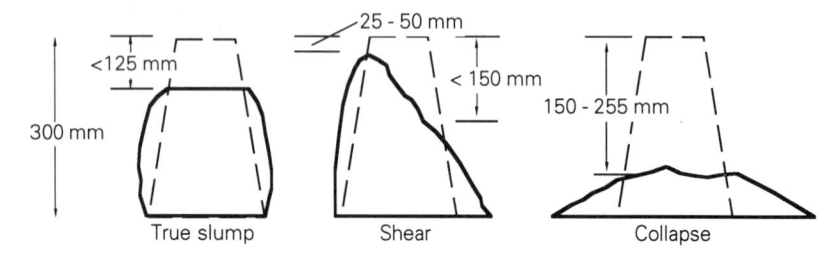

Figure 5.4 Slump – true, shear and collapse

repeated. If the shear slump persists, then this can be taken as an indication of lack of cohesion of the mix and is an indication that the mix is harsh or lean (Neville and Brooks, 1990, p. 84).

In general the slump is proportional to the water content squared.

The results are useful as an indication of consistency only when the slump is between 20 and 125 mm. The slump is normally about 80 mm maximum. If the slump is less than 20 mm, then other methods of measurement have to be used, e.g.

Compaction factor test	AS 1012.3 (Method 2)
Vebe test	AS 1012.3 (Method 3)
Flow tube test	ASTM C 124

A slump of (> 125 mm) is difficult to measure using the conventional slump test. With flowing concrete there is a good chance that if a slump test were used, the concrete would overflow the boundaries of the bottom plate and the results would be invalid. Therefore a flow table test is used for these concretes. The procedure is similar to that outlined above for the slump test, except that the cone is turned out on a table, as shown in Figure 5.5.

The table is made of a heavy steel plate. After the removal of the cone, the table is tilted and then allowed to drop. This is done 15 times, each cycle taking about 4 seconds. The maximum spread of the concrete parallel to the two sides of the table is measured, and the average of these two values, to the nearest millimetre is taken to represent the flow of the concrete. A value of 400 indicated medium workability, whereas a value of 500 represents high workability (Neville and Brooks, 1990, p. 90). A major drawback of this test is that the frame supporting the plate has to be permanently anchored to the floor, the weight of the apparatus is about 95 kg, and the operator needs to have strength and wear ear plugs. Thus the apparatus is not easily transportable to a building site for routine tests (Mor and Ravina, 1986).

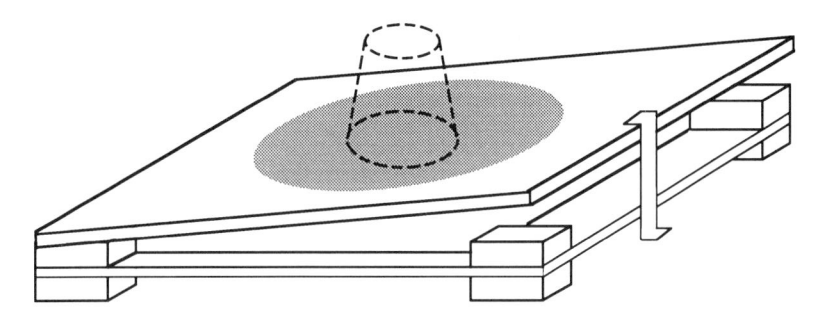

Figure 5.5 Flow table

Table 5.3 Workability, slump and typical uses of concrete

Degree of workability	Slump (mm)	Typical use for concrete
Very low	<25	Roads (power-operated vibrators needed for compaction)
Low	25–50	Mass concrete foundations without vibration
Medium	50–100	Flat slabs
High	100–175	Congested reinforcement

A portable simplified version of the flow table is described by the German standard DIN 1048. This is a smaller apparatus, and has been shown by Mor and Ravina (1986) to give results that correlate well with the slump test, and while they agree that the results in the field might be operator dependent, they consider that the flow table should be used to test the new more fluid range of concretes.

The order of magnitude of the slump for different workabilities is shown in Table 5.3.

It should be noted that the slump is not directly proportional to the workability, since the nature of the aggregates will affect this, but it is a useful on-site test to check for any variations in the concrete supplied.

QUESTIONS

1. List the major factors that determine the compressive strength of set concrete.
2. Why is the early strength of concrete (up to about 90 days) proportional to the C_3S content of the cement, and why does it change after this time?
3. What are the essential functions of cement and aggregate in concrete?
4. Why should aggregate materials used in the production of ordinary concrete be non-porous? What effects have to be taken into account if porous aggregates are used in the production of lightweight concrete?
5. Describe why, if C_2S is a component of cement, its presence is undesirable in blast furnace slag aggregate.
6. Why is the maximum particle size of aggregate that can be used in a mix dependent on the proposed use of the concrete?
7. Why are aggregates graded for use in concrete?
8. What is the alkali/aggregate reaction?
9. Why are clays often added to mortars, and yet they are classed as harmful impurities when present with aggregates used in concrete?

10. What is the essential difference between the 'laws' proposed by Abrams and Feret to describe the dependence of the compressive strength of set concrete on the water/cement ratio used in the concrete? Sketch a graph to illustrate your answer.
11. Describe the origins of voids in normal strength concrete and explain how these affect the strength of concrete.
12. Define workability and describe the factors that can affect the workability of a mix.
13. Is there any relationship between workability and the strength of set concrete? Give reasons for your answer.
14. What is the 'consistency' of concrete? Describe the essential features of the compaction factor test, the Vebe test and the flow tube test for the determination of consistency of fresh concrete.

References

Mindess, S. (1989) Interfaces in concrete. In J. Skalny (ed.), *Materials Science of Concrete I*, The American Ceramic Society, Westerville, OH, pp. 163–80.

Mor, A. and Ravina, D. (1986) The DIN flow table, *Concrete International*, **8** (12), 53–56.

Neville, A. M. (1981), *Properties of Concrete*, 3rd edn, Pitman, London.

Nagarajan R. and Antill J. M. (1978) *Australian Concrete Inspection Manual*, Pitman Australia, Carlton, p. 187.

Neville, A. M. and Brooks, J. J. (1990) *Concrete Technology*, Longman Scientific & Technical, Essex, UK.

Popovics, S. (1994) The slump test is useless – or is it?, *Concrete International*, **16** (9), 30–33.

6 | Standard tests for cements, cement pastes, mortars and concrete

The physical and chemical properties of commercial cements are specified by national standards and they are measured by standard tests. It is unfortunate that different countries have different standards and different specifications for their cements. This means that the results obtained from the different national standard tests are often not directly comparable one to the other, and they are frequently quite arbitrary.

6.1 CEMENTS

6.1.1 Chemical and mineralogical composition

Detailed methods that can be used to measure the chemical composition of Portland cement are given in the standards ASTM C14–88 and AS 2350.2. The chemical composition of a cement is normally measured in order to check whether the levels of impurities in the cement are within prescribed levels. Such tests are carried out for MgO, SO_3, Na_2O and K_2O. Chemical analysis of CaO, SiO_2, Al_2O_3 and Fe_2O_3 can be used to calculate the hypothetical amounts of the phases present using the Bogue equations. As was pointed out in Chapter 3, there is wide variation in the amounts of the phases in the cements, and maximum or minimum limits for the **C_2S, C_3S, C_3A** and **C_4AF** in a cement are only set when special properties are required for example for the production of low heat or sulphate resistant cements (Mindess, 1983). Such specifications are set out in the Australian standard AS 3972-1991 (the corresponding American standard is ASTM C150-92).

6.1.2 Fineness

The fineness of a cement is determined by its particle size and the particle size distribution. These have a large effect on the setting properties of the cement because this takes place by reactions between the surface of the cement particles and water. The larger the surface area, the greater the rate of reaction. Normally only the specific surface area of the cement is measured, even though it is recognized that the particle size distribution is also of importance and two batches of cement with the same surface area could have different particle size distributions and therefore different reactivities.

Of the many methods that can be used to measure the surface area of cement the two most common use the sedimentation behaviour of the particles in a viscous medium and the permeability of a bed of the cement particles to air. In the first method a Wagner turbidimeter might be used, as described in ASTM C115–93, while the second uses the Blaine method, described in ASTM C204–92 and AS 2350.8). It must be borne in mind that these methods give only comparative results, and the absolute values of surface area can be very different from those determined by these measurements. A full description of these test methods is given by Lea (1970, pp. 370–87).

6.2 CEMENT PASTES

6.2.1 Plasticity

The 'plasticity' of a cement paste is defined by its normal consistency and its initial and final set. A paste is made from the cement and a defined amount of water and is then placed in the Vicat apparatus. This apparatus measures the depth of penetration of a 1 mm diameter needle under a 300 g load into the cement paste (Lea, 1970, pp. 362–4). The initial set penetration is taken when the paste has started to stiffen, and the final set is when the paste has hardened sufficiently to sustain a load. There is wide variation between the requirements of different countries, as shown in Table 6.1, even though the same Vicat test apparatus is used to determine the initial and final set of the cement paste (as described in ASTM C191–92 or AS 2350.4).

Table 6.1 Standards for minimum and maximum setting times

Standard	Minimum time for the initial set	Maximum time for the final set
AS (Australia)	45 minutes	10 hours
BSI (UK)	45 minutes	10 hours
German	60 minutes	12 hours
ASTM (USA)	45 minutes	6 hours 15 min.

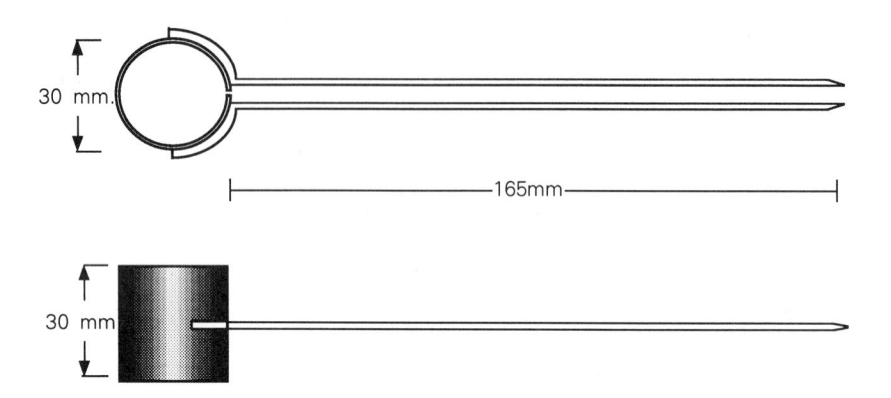

Figure 6.1 Le Châtelier apparatus

Temperature has an effect on the rate of setting of a cement. In general the rate will increase as the temperature is increased. The amount by which rate increases is dependent on the type of cement.

6.2.2 Unsoundness

The term 'unsoundness' refers to expansion of the cement that occurs after the final set (Mindess, 1983). This expansion can occur by the hydration of MgO or free lime (CaO), or by sulphate attack. Since these reactions are slow, it is necessary to carry out accelerated tests on the cement pastes at high temperatures (100 °C or higher). One such test is the Le Châtelier test (used only for free lime) in which the cement paste is put into a split brass cylinder as shown in Figure 6.1. After the paste has set, the cylinder is place in boiling water. Any expansion due to hydration will be shown by an increase in the separation of the indicator pointers which are attached to the cylinder. Full details of this test can be found in ASTM C151 or AS 2350.5.

More details of this test and also of other test used to measure unsoundness due to the presence of magnesium hydroxide or sulphate attack are given by Lea (1970, pp. 366–370).

6.3 MORTAR

The properties of mortars not only depend on the type of cement used, but also on the sand. The various standards all specify 'standard' sand, but this will vary from one country to another. Thus the results obtained with the tests will not be comparable since the variation in the sand will cause variation in the results. Some of the tests that are done include measurement of the consistency, water retention and strength (compressive, flexural and tensile).

6.3.1 Consistency

The flow index of a mortar is used as a measure of its consistency and is determined by measuring the increase in diameter of a cone of mortar after a specified jolting on a flow table (Mindess, 1983). The ratio of the final diameter of the cone to the initial is the 'flow index' (Lenczner, 1972).

6.3.2 Water retention

The water retention of a wet mortar can be determined by the amount of water that can be removed from the mortar by the application of a number of layers of blotting paper weighed down on the mortar by a standard weight (Lenczner, 1972).

6.3.3 Compressive strength

The compressive strength is simple to measure, and is normally carried out on cubes of set mortar, but, as in the previous cases, the test specimens and test conditions vary between standards. Not only do the dimensions of the test specimens vary, but also the type of sand used, the sand/cement ratio and the water/cement ratio. The time of curing before the measurements are made can be 7, 14 or 28 days. ASTM C109–92 requires 50 mm cubes for this test, whereas AS 2350.11 requires 40×40×160 mm prisms.

6.3.4 Flexural strength

Similar mixes to those used for the measurement of compressive strength are used to make up bars to measure the flexural strength in three or four point bending. The ASTM C348–93 standard is for three point bending using 40×40×160 mm prisms.

6.3.5 Tensile strength

Tension tests are sometimes done on mortars. The mortar is cast into a tensile test specimen shown below. Standards exist which specify the type and amount of sand to be used, the details of curing etc., but there is still normally a large scatter in the results obtained by this test (Neville, 1981, p. 55). The ASTM C190–85 standard for this test was discontinued in 1991.

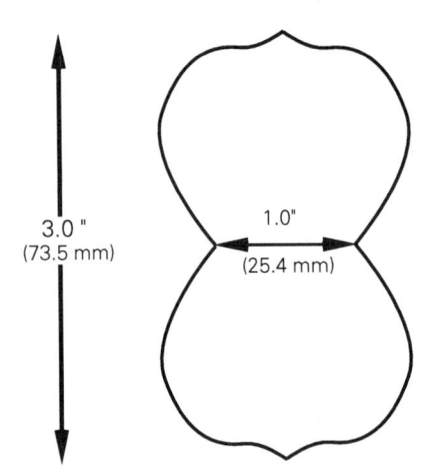

Figure 6.2 Tensile test specimen

6.4 CONCRETE

6.4.1 Fresh concrete

The slump test (AS 1012.3 or ASTM 143–90) has already been described in Chapter 5. Other tests described in the Australian standard include the compaction factor test, the Vebe test and the compatibility index of fresh concrete. Air entrainment can be determined by measuring: the reduction in concrete volume with increased air pressure (ASTM C231–91b) or the air volume released when concrete is dispersed in water (ASTM C173–93). All these tests are also set out in AS 1012.4-1983.

ASTM C138–92 or AS 1012.5-1983 gives the method for the determination of mass per unit volume of freshly mixed concrete and the bleeding of concrete can be assessed by ASTM C232–92 or AS 1012.6-1983.

The setting times are also important factors for fresh concrete and its initial and final set is determined by the use of the Vicat needle in a similar manner to that of cement paste. The test is described fully in ASTM C403.

6.4.2 Set concrete

(a) Drying shrinkage and creep

The drying shrinkage of concrete can be measured by tests described in AS 1012.13-1970; the creep is determined by the measurement of the change in length of concrete cylinders in compression (ASTM 512–87 or AS 1012.16-1974).

(b) Compressive strength

The compressive strength of concrete is one of its most important properties. This is because in most structural applications concrete is employed primarily to resist compressive stresses. In cases where strength in tension or in sheer is of primary importance, the tensile or sheer strengths are often estimated from its compressive strength (Neville, 1981, Ch. 8). The compressive strength is therefore used as a measure of the over-all quality of the concrete and thus as an indication of the other properties related to deformation or durability.

It is very easy to conduct a compressive strength test on concrete. Unfortunately, the standard test methods that are used to measure this property vary widely from one country to another. The test is carried out on cubic test specimens in most of Europe whereas cylindrical test specimens are used in the USA, France, Canada, Australia and New Zealand. The dimensions of the specimens and details of the test methods used also differ, for example the loading rate and the methods of capping the samples to obtain an even load distribution. There is the additional problem of variation in the methods used to compact the samples. The measured strengths will also depend on the method of curing, curing time, curing temperature, type of cement used etc. The most commonly quoted strength is the '28-day' strength, although sometimes a '7-day' strength is used in quality control. The assumption is made that the strength will continue to increase after 7 days, but with modern concrete formulations this is not always the case.

The Australian standard (AS 1012.8-1986) describes two test specimens for the compression test and also the indirect tensile test (Brazilian test). These are cylinders 150 mm diameter and 300 mm in height (for aggregates <40 mm) and also 100 mm diameter and 200 mm high (for aggregates <20 mm). It is pointed out that the data obtained from the two specimens shall not be combined. Standard methods for compaction by rodding, vibration and ramming are described. The methods used for curing are dependent on the location of the test within Australia. Two different zones are defined which are the Standard Tropical Zone (Queensland, Northern Territory and Western Australia, north of latitude 25 °S), and the Standard Temperate Zone, which covers the rest of the country.

Samples for compressive strength measurements may also be prepared by removing cores from concrete structures, this is often done to check on 'suspect' concrete. In this case there is difficulty in obtaining cores of the standard size in particularly the 2:1 length/diameter ratio required by some standards. Empirical correction factors have to be applied for any difference in length/diameter ratio and differences in diameter of the cores from those specified by the standards.

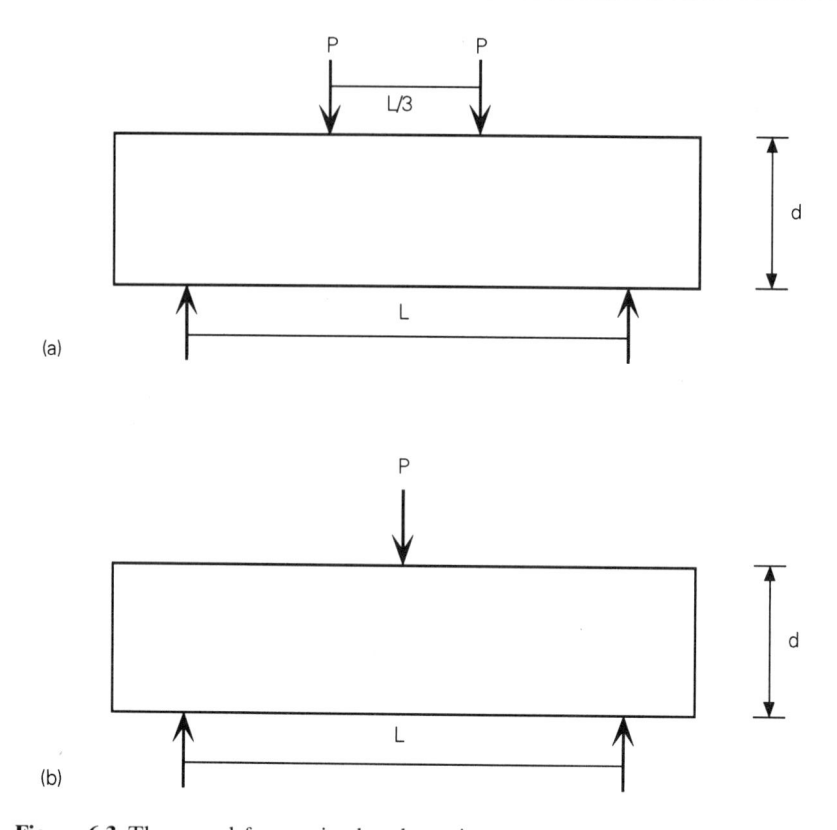

Figure 6.3 Three and four point bend specimens

(c) Tensile strength

(i) Flexure test
Most flexure tests are carried out in four- or three-point bending of a beam of concrete having rectangular cross section.

The test for four-point bending is described in AS 1012.11-1985 and ASTM CC78-84, and that for three-point bending in ASTM C293-79.

For a four-point bend test, the Modulus of Rupture (R) is calculated from the relationship

$$R = PL/bd^2$$

where

R = Modulus of Rupture (MPa),

P = maximum total load on beam (N)

L = span of the beam (mm),

b = cross-section width of the beam (mm), and

d = cross-section depth (mm)

For a three-point bend test the Modulus of Rupture is given by

$$R = 3PL/2bd^2$$

In the Australian standard specifications for specimens for the flexure test are given only for four-point bending. Two test specimens are described, one with cross-section 150×150 mm, length at least 500 mm, and maximum aggregate size 40 mm, and the other 100×100 mm in cross section, length at least 350 mm, and maximum aggregate size 20 mm. It is emphasized that the data obtained from the different test specimens must not be combined. The compaction and curing conditions are similar to those described for the compression or indirect tensile test specimens.

In the American standards the size of the test beams are not specified, except that the depth of the beams should be equal to L/3.

(ii) Indirect tensile strength (Brazilian test)

In this test tensile failure is by splitting of a specimen that is being compressed. A cylinder or disc of concrete is loaded across the diameter of the sample. Failure in tension occurs with the test cylinder splitting vertically down its diameter as shown in Figure 6.4.

The tensile strength (σ_t in MPa) is then given by

$$\sigma_t = 2P/\pi Ld$$

where

P = maximum applied force (N) on the cylinder,

L = length (mm),

d = diameter (mm).

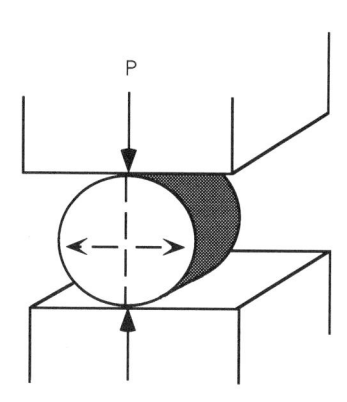

Figure 6.4 Brazilian test specimen

The two Australian standard test specimens have the same specification as for the compressive strength test. Details of the test method to be used is given in AS 1012.10-1985. The load is spread on the diameter of the samples by the interposition between the compression machine anvils and the specimen of two pieces of hardboard 5 mm thick and 25 mm wide and at least as long as the specimen. No standard size for the test specimens are defined in ASTM C496-90.

6.5 NON-DESTRUCTIVE TESTS

The non-destructive testing of concrete is becoming increasingly important as a means of assessing existing structures before their failure and their subsequent expensive repairs or replacement. There is also a need to be able to monitor the structures built with newly developed concrete materials for which long-term performance data is not available.

The problem with obtaining quantitative measurements of material properties of concrete is that concrete normally contains microcracks, voids and other flaws. This means that the errors in the test results can be extremely large unless a very large number of measurements are taken, so that statistics can be used to analyse the results. The use of destructive methods necessitates the preparation and testing of a very large number of test samples in order to obtain reliable data, therefore non-destructive test methods have the great advantage that a large number of measurements can be done on the one sample. Some of the various methods that can be used for the determination of material properties of concrete by non-destructive means are described briefly below. A comprehensive description of the non-destructive testing of concrete can be found in Orchard (1979) and more recently, Malhotra and Carino (1991).

6.5.1 Cover depth of concrete over reinforcement

The magnetic field generated at the surface of reinforced concrete is distorted by the reinforcing bar within the concrete and the magnitude of the distortion can be related to the depth of concrete covering the reinforcement (Mallett, 1994). A wide range of these instruments ('covermeters') are available but most require a knowledge of the diameter of the rebar, so the calibration of the instruments needs to be checked by drilling.

6.5.2 Delaminations in concrete

Delaminations in concrete are normally the result of corrosion of reinforcement. These can be detected by impulse radar and by infra-red thermography. The use of radar is based on reflections of electromagnetic pulses from discontinuities within the structure. Infra-red thermography is used to measure spatial differences in the surface temperature of a structure. The extra impedance to heat flow caused by a delamination within the concrete will alter the local heat flow and hence the surface temperature (Manning, 1992).

6.5.3 Elastic modulus

The dynamic elastic modulus can be determined using vibration resonance techniques. The apparatus used is described in ASTM C215-91. This test is carried out in the laboratory on either cylindrical or rectangular beams. Depending on the location of the vibrator and pick-up, different modes of vibration can be used: longitudinal, flexural or torsional. The resonant frequency of the test sample can be measured and this can be related to the elastic modulus of the sample. Details of this test are give by Jones (1962). One application for this test is the detection of deterioration of concrete in freeze/thaw conditions (ASTM C666-92).

Vibrational techniques can be used 'in the field' by the generation of surface waves. The pulsed vibrations are produced by a mechanical hammer, the velocity of the pulse is measured and used to determine the dynamic modulus of elasticity of the sample.

Ultrasonic methods can also be used for the determination of Young's Modulus, as described in ASTM C597-83 (Reapproved 1991). Piezo-electric crystals are used to generate and detect the ultrasonic waves. This method can only be used in situ if access can be gained to both sides of the concrete.

6.5.4 Density

The measurement of the density of concrete in situ can be done using the absorption of γ radiation. The amount of the radiation being back scattered or transmitted is dependent on the density of the concrete. γ-ray sources of between 0.1 and 1 MeV have to be used to ensure that only Compton scattering occurs and prevent pair production and photo-electric absorption (a Caesium-137 source is used in the ASTM C1040–93 standard). Detection of the γ-rays is done using a sodium iodide scintillation counter, Geiger-Müller counter or proportional counter.

Neutron absorption can also be used to measure density and moisture content of the concrete.

6.5.5 Hardness

The measurement of hardness in situ is normally carried out using the Schmidt rebound hammer. A plunger is held in contact with the concrete,

Table 6.2 Methods of measuring hardness of set concrete

	Indentation	Rebound	Pull-out	Windsor probe
Parameter measured	Dimensions of an indentation	Rebound of a hammer or pendulum	Force to pull out	Depth of penetration

an impact force is applied and the rebound measured. The use of the rebound hammer is fully described in ASTM C805.

The various methods of measuring the hardness and similar properties are shown in Table 6.2. They can be applied to any concrete in situ and have the advantage that the tests are simple, but the results show large variation and only the surface or near surface properties are measured. The harness results give an indication measure of the strength of the concrete, but correlating calibrations have to be carried out.

QUESTIONS

1. Why are chemical analyses of cements carried out? Why is there a wide variation in the calculated phase composition of any one type of cement? Would you expect the calculated phase composition to be the same as the true phase composition? If not, why not?
2. Outline the methods that are used to measure the surface area of cement. What effect does the surface area have on the properties of cement?
3. Two cements, with identical surface area and composition, exhibit different reactivities. What other measurements are needed to explain this?
4. What is unsoundness? How can unsoundness due to magnesium hydroxide or sulphate attack be measured?
5. If you were comparing the compressive strength of a concrete that you have made with values obtained from other researchers in different countries, what factors should you bear in mind?
6. What are the advantages of non-destructive testing of concrete?

References

Carino, N. J. (1994) NDT methods for flaw detection in concrete. In P. W. Brown (ed.), *Cement: Manufacture and Use*, American Society of Civil Engineers, New York, pp. 175–92.

Jones, R. (1962) *Non-Destructive Testing of Concrete*, Cambridge University Press, Cambridge.

REFERENCES

Lea, F. M. (1970) *The Chemistry of Cement and Concrete*, 3rd edn, Edward Arnold, London.

Lenczner, D. (1972) *Elements of Load Bearing Brickwork*, Pergamon, Oxford, Ch. 2.

Malhotra, V. M. and Carino, N. J. (1991) *Handbook on Non destructive Testing of Concrete*, CRC Press, Boca Raton.

Mallett, G. P. (1994) *Repair of Concrete Bridges*, Thomas Telford, London, p. 74.

Manning, D. G. (1992) Design life of concrete highway structures – the North American scene. In G. Somerville (ed.), *The Design Life of Structures*, Blackie Academic, Glasgow, pp.144–53.

Mindess, S. (1983) Concrete Materials. *Journal of Materials Education*, **4**, 982–1046.

Neville, A. M. (1981) *Properties of Concrete*, 3rd edn, Pitman, London.

Orchard, D. F. (1979) *Concrete Technology,* Vol. 2, 4th edn, Applied Science Publishers, London, Ch. 4.

7 Some additives (admixtures) used in mortar and concrete

An admixture is defined by the American Concrete Institute (ACI) as 'a material other than water, aggregates, hydraulic cement and fibre reinforcement, used as an ingredient of concrete or mortar, and added to the batch immediately before or during its mixing' (ACI Committee 212, 1989). The use of admixtures in concrete is certainly not new, in Roman times blood or eggs were added to concrete to improve its properties and in the thirteenth-century linseed oil was added to make the concrete water tight (Aïtcin and Baalbaki, 1995).

According to the report by the ACI Committee 212 (1989), the admixtures may be added:

- to modify the properties of fresh concrete, mortar or grout such as to:
  Increase workability without increasing water content or decrease the water content at the same workability.
  Retard or accelerate time of initial set.
  Reduce or prevent settlement or create slight expansion.
  Modify the rate and/or capacity for bleeding.
  Reduce segregation.
  Improve pumpability.
  Reduce the rate of slump loss.

- to modify properties of hardened concrete, mortar and grout so as to:
  Retard or reduce heat evolution during early hardening.
  Accelerate the rate of strength development at early ages.
  Increase strength (compressive, tensile or flexural).
  Increase durability or resistance to severe conditions of exposure, including application of deicing salts.
  Decrease permeability of concrete.
  Control expansion caused by the reaction of alkalies with certain aggregate constituents.

ACCELERATORS, RETARDANTS AND STABILIZERS

Increase bond of concrete-to-steel reinforcement.
Increase bond between existing and new concrete.
Improve impact resistance and abrasion resistance.
Inhibit corrosion of embedded metal.
Produce coloured concrete or mortar.

As pointed out by Aïtcin and Baalbaki (1995) the terminology used to describe the various admixtures that can be added to concrete is extensive and can create confusion rather than clarification. They liken admixture technology to alchemy rather than physics/chemistry. Each country has its own terminology and the standards vary; for example, the ASTM C494 describes seven types of water reducers and the European standard EN934-2 lists three.

In this chapter the following additives and their effects on concrete are discussed:

- accelerating and retarding additives that can be used to control the rate of setting of concrete;
- air entrainment additives that increase workability of fresh concrete and produce a frost resistant concrete;
- materials that exhibit pozzolanic activity (such as pozzolanas and blast furnace slag) affecting the rate of heat evolution of setting concrete and the durability of set concrete;
- additives that increase the flow and workability of fresh concrete without any increase of the water/cement ratio.

The use of other additives than those in the above list and their effects on concrete will be discussed in later chapters.

7.1 ACCELERATORS, RETARDANTS AND STABILIZERS

The effect of the addition of gypsum to Portland cement to prevent a flash set has been discussed in the previous chapters. There are many other salts and organic compounds that are added to cement to control the rate of setting of the cement. Some salts retard the set, others accelerate it and some retard the set when present in small amounts but accelerate the set when present in larger amounts.

7.1.1 Accelerators

Many inorganic and organic compounds can be used to accelerate the set of concrete. The two most important salts which are added to cement are calcium chloride and sodium chloride. These are often used in concrete laid in cold weather to reduce the curing and setting time so that formwork can be removed within a reasonable time. Additions of calcium chloride < 1% retards the set, but larger amounts accelerate the set, and additions of > 3% often result in a flash set (Lea, 1970, p. 301). However the addition of calcium chloride additions to prestressed concrete and reinforced concrete that is exposed to moist conditions in order to accelerate the set is now not recommended in the USA and Australia. This is because the chloride ions can cause pitting corrosion of steel used to prestress or reinforce the concrete. Calcium chloride additions also decrease the resistance of the concrete to sulphate attack (ACI Committee 212, 1989). There are non- corrosive accelerants available, such as certain nitrates, formates and nitrites, but these tend to be less effective and more expensive than calcium chloride.

7.1.2 Retarders

Retardation of the setting of concrete may be brought about by the addition of organic compounds which contain CHOH groups (e.g. sugars, starch and derivatives of cellulose compounds). These retard the set by the adsorption of the molecules onto the surface of the cement particles (Lea, 1970, p.306). About 0.05 mass% sugar will retard the setting time for about four hours and a 1% solution almost completely inhibits real setting and hardening, though there may be an immediate rapid stiffening giving the appearance of a quick set. This complete inhibition of setting is used in the case of malfunction of concrete mixers to prevent the setting of the concrete in the mixer.

Sugars, and other organic compounds, are often added to cements that are used to bond the steel casings of oil wells to the rock formations and to seal off porous rock through which the drill passes. The cement has to be able to be pumped into position, often to a great depth and is then subject to high temperatures and pressures. It is a requirement that the cement remain fluid for a period of several hours, and then harden fairly rapidly. Other retardants include soluble zinc salts, soluble borates and phosphates or hydrocarboxylic acids (citric, tartaric, gluconic) and their salts.

7.1.3 Stabilizers

Another group of chemicals that are used to control the set of concrete are known as stabilizers. These act to completely prevent the hydration of both the calcium silicates and calcium aluminate, unlike retardants which do not prevent the hydration of the calcium aluminate. These materials have been used to minimize the waste of excess ready mixed concrete which is returned from the job-site to the supplier for disposal (Attiogbe and Farzam, 1994). The stabilizers can prevent the hydration reactions for up to 72 hours. The concrete may then be used after the addition of an activator which breaks down the protective layer around the cement particles and the concrete then sets in the normal manner. This system can also be used for the utilization of left-over concrete in pump lines and also for concrete that has to be transported over long distances. Unfortunately, Attiogbe and Farzam (1994), do not identify the chemicals used as stabilizers or activators.

7.2 ADMIXTURES FOR AIR ENTRAINMENT

Entrained air increases the workability of concrete, since it increases the paste volume and the air bubbles act as a lubricant between the solid particles. Additives are often used to promote entrainment of air in order to obtain this increase in workability, but more importantly they are used to entrain air to increase the durability of the concrete in climates where it undergoes freeze/thaw cycles (see Chapter 12). Segregation and bleeding are also reduced by air entrainment and this helps to prevent the entrapment of bleed water under either the reinforcement or large aggregate particles, which could result in corrosion of the reinforcement or a reduction in the strength of the concrete. (Segregation and bleeding are discussed in Chapter 9.)

The deliberately entrained air bubbles are uniformly distributed in the concrete and are very much smaller than the accidentally entrapped air voids in concrete, which are of random size and distribution. There are more than 10^9 bubbles per cubic yard of air entrained concrete (ACI Committee 212, 1989). The cement paste is normally protected from damage due to freeze/thaw cycles if the minute air bubbles are less than 200 μm apart.

The Romans used air entrainment additives such as ox-blood in order to increase the durability of the concrete. The modern usage of air entrainers stems from the accidental observation that some sections of concrete roads in the USA showed increased frost durability, and these were found to be made from cement which had been contaminated by beef tallow which had been used as a grinding aid.

Various chemicals are now used as air entrainment agents, these include alkali salts of wood resins (sodium abietate), alkyl or aryl sulphonates, alkyl sulphates, and salts of fatty acids derived from animal and vegetable fats and oils.

The entrainment of air does reduce the strength of the concrete, but because of its increased workability, a lower water/cement ratio can be used which increases the strength enough to offset the effects of the entrained air. Air entraining agents are often added to high-performance concrete, which has a very low water/cement ratio, in order to increase

| 86 | SOME ADDITIVES (ADMIXTURES) USED IN MORTAR |

ANIONIC AIR-ENTRAINING AGENT

Negatively charged head ⊖⌇ Non-polar tail

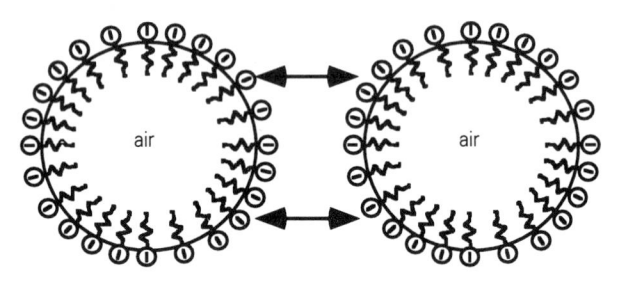

Repulsion between negative charges on the
surface of the air bubbles prevent coalescence.

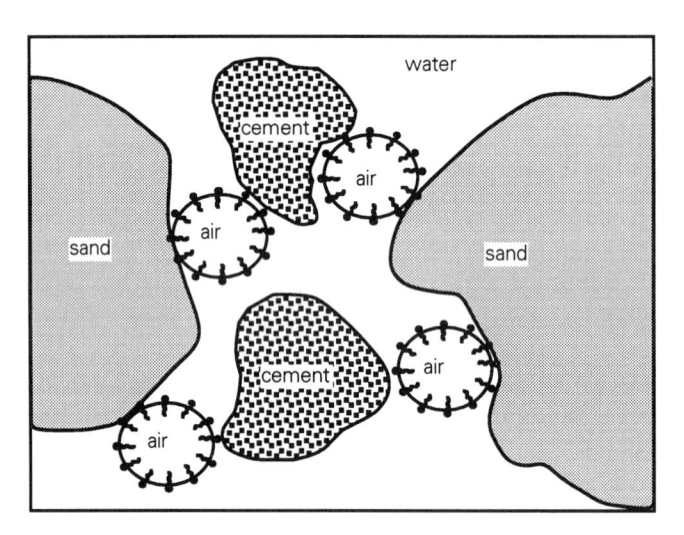

Figure 7.1 Mechanism by which air entrainment agents work in concrete (redrawn from Williams and Swail, 1988)

its workability (Aïtcin, and Baalbaki, 1995). (The production of high-performance concrete is described in Chapter 8.)

The mechanism by which air entrainment agents work is complex, but most agents contain hydrophobic and hydrophilic groups at opposite ends of a carbon chain. These stabilize the entrained air and prevent the air bubbles coalescing, as shown in Figure 7.1.

7.3 ADDITIONS OF POZZOLANAS

Pozzolanas are defined as siliceous or siliceous and aluminous materials which in themselves possess little or no cementitious value but will, in finely divided form and in the presence of moisture, chemically react with calcium hydroxide at ordinary temperatures to form compounds possessing cementitious properties (ASTM C618-93). Pozzolanas include naturally occurring materials (mainly volcanic in origin), and artificial materials such as fly ash, calcined rice hulls and blast furnace slag. Although some fly ashes and blast furnace slags possess cementitious value and so do not strictly conform to the definition of pozzolanas, they are included in this section as they react in a manner similar to true pozzolanas.

Pozzolana based materials are added to Portland cement based concretes and mortars in order to reduce the rate of evolution of heat during setting, reduce the cost of the concrete and to improve durability. The pozzolanic materials used can be divided into two classes, natural and artificial.

7.3.1 Natural pozzolanas

According to the ACI Committee 232 (1994) a 'Natural Pozzolan' is defined as 'either a raw or calcined material that has pozzolanic properties (e.g. volcanic ash or pumicide, opaline chert or shales, tuffs and some diatomaceous earths)'.

The ASTM–C618 definition of a natural pozzolana is

> Raw or calcined natural pozzolans that comply with the applicable requirements for the class given herein, such as some diatomaceous earths; opaline cherts or shales; tuffs or volcanic ashes or pumicites, any of which may or may not be processed by calcination; and various materials requiring calcination to induce satisfactory properties such as clays and shales.

The natural pozzolanas of volcanic origin are unconsolidated volcanic ash or ash which has been compacted by overlaying rock (tuffstein). This rock, after grinding, produces trass.

Pozzolanic properties are also shown by diatomaceous earths, which consist of the porous siliceous skeletons of microscopic organisms that lived in either fresh or sea-water.

Burnt clays with pozzolanic activity are made by heating (burning) clays or shales to above 600 °C. The exact temperature needed is dependent of the nature of the clay and the method of heating. It is necessary to grind the burnt clay into a fine powder before it can be used as a pozzolanic material.

Table 7.1 Composition (mass%) of some natural pozzolanas (after Lea, 1970, p.415)

	SiO_2	Al_2O_3	Fe_2O_3	CaO
Rhenish Trass (Germany)	55	16	4	4
Santorian Earth (Crete)	63	13	5	4
Baculi (Naples)	60	19	5	5
Rumanian Trass	63	12	2	7
Pumicite (USA)	72	13	1	1
Burnt Clay	58	18	9	3

(Plus minor amounts of TiO_2, MgO, Na_2O, K_2O etc.)

The main common feature of the natural pozzolanas is that they all are silicates and have a high specific surface area. Most of these silicates are glassy, but some are crystalline. The ash pozzolanas are glassy silicates formed by the rapid cooling of molten droplets of lava, and usually consist of porous spherical particles. The older tuffstein deposits have often been altered by the action of superheated steam and carbon dioxide below the earth's surface to form crystalline zeolites. Zeolites are framework silicates which contain large channels in their crystal structure which result in them having a large specific surface area. (Zeolites are also often used as ion exchange media to soften water.)

Some typical compositions of natural pozzolanas are shown in Table 7.1.

7.3.2 Artificial pozzolanas

Artificial pozzolanic materials include fly ash, calcined rice hulls and blast furnace slag.

(a) Fly ash

Fly ash is produced as a by-product of coal combustion in thermal power stations. Large amounts of fly ash are produced annually, for example, in 1989 the world production was in the order of 400 million tonnes (Malhotra, 1993). The fly ash consists of fine spherical particles which contain 60–90% glassy component. The composition of the fly ash is very variable, but in general contains 45–48 wt% silica and 18–28 wt% alumina, together with smaller amounts of iron oxides, calcia, magnesia etc. Most ash is low calcium fly ash (ASTM Class F), and this material requires the presence of calcium hydroxide to form compounds with cementitious properties. It is normally a product of the combustion of anthracite or bituminous coal. A high calcium fly ash (ASTM Class C) is produced by the burning of lignite (brown coal) or sub-bituminous coals (Alonso and Wesche, 1991). The class C fly ash has cementitious properties (the calcium hydroxide necessary to activate the pozzolanic reactions can be produced from the fly ash itself) as well as pozzolanic properties (Malhotra, 1993).

Alonso and Wesche (1991) describe the sub-division of fly ashes into four categories according to chemical composition. Types I and II correspond approximately to the ASTM classes F and C respectively and contain > 35 wt% SiO_2. Types III and IV fly ashes contain < 35 wt% SiO_2 combined with very high amounts of CaO. These are almost inactive as pozzolanic materials, are not suitable for use in concrete and may cause unsoundness in concrete. The fly ash will also contain unburned coal particles, the amount of which is dependent on the details of the combustion processes of the coal. The carbon content is important as the carbon particles have high surface area and have high absorbency of water and organic or inorganic admixtures that might be added to the concrete. In general, the pozzolanic activity of fly ash increases with increasing glass content and with decreasing particle size and carbon content.

(b) Rice hulls

Rice hulls can also be used to make a pozzolanic material. About 100 million tonnes of rice hulls are produced each year in the world. Combustion of these rice hulls burns out the carbon and leaves a silica residue. About 200 kg of silica ash is produced from each tonne of husks. However the ash that is produced by burning is normally crystalline and has low pozzolanic activity, but if the burning temperature is carefully controlled to between 500 and 700 °C, a high specific surface area pozzolanic material is formed (Malhotra, 1993).

(c) Blast furnace slag

The potential hydraulic setting of glassy blast furnace slag has been known since 1862, when it was discovered by Emil Langdon, and admixtures of blast furnace slag and Portland cement have been produced commercially for about 100 years (Reeves, 1985).

Blast furnace slag is a by-product of the production of pig iron. The function of the slag in the blast furnace is to remove impurities which originate in the raw materials (iron ore, limestone and coke). The main use of the slag in the iron making process is in the removal of sulphur, and the composition of the slag has to be adjusted according to the impurities present in the raw materials. Almost equal quantities of slag and iron are produced. Blast furnace slag can either be used as a raw material for the production of Portland cement clinker, or as a partial substitute for Portland cement in concrete.

SOME ADDITIVES (ADMIXTURES) USED IN MORTAR

Table 7.2 Composition (mass%) of blast furnace slag and Portland cement clinker

	CaO	SiO$_2$	Al$_2$O$_3$	MgO	Fe$_2$O$_3$	
Blast furnace slag	30–50	28–38	8–24	1–18	1–3	plus MnO, S.
Portland cement	60–67	17–25	3–8	<3	0.5–6	plus Na$_2$O etc.

The composition of slag varies widely, due to variations in the compositions of the ore, flux, coke etc. Typical slag composition, compared with Portland cement, is shown in Table 7.2.

For use with Portland cement in concrete, the molten slag is rapidly cooled by pouring it into a large excess of water, or by quenching it with water jets. The aim of the rapid cooling is twofold, firstly to produce fine particles, and secondly to prevent crystallization of the slag.

A pelletizer has been developed in Canada which is capable of producing several different types of slag products. The advantage of this machine is that it uses much less water than previous processes (1 m^3/t slag, as compared with 100 m^3/t slag). The fractions obtained are partly crystalline porous nodules (4–15 mm) which are suitable for lightweight aggregates (pellets) and granules of < 4 mm which are glassy and which can be added to Portland cement as a pozzolanic material. A diagram of the process is shown in Figure 7.2 (redrawn from Regourd, 1983). The drum rotates at 300 rpm.

Water-cooled slag is partially substituted for Portland cement in a manner similar to that of pozzolanas. The slag can either be dried and

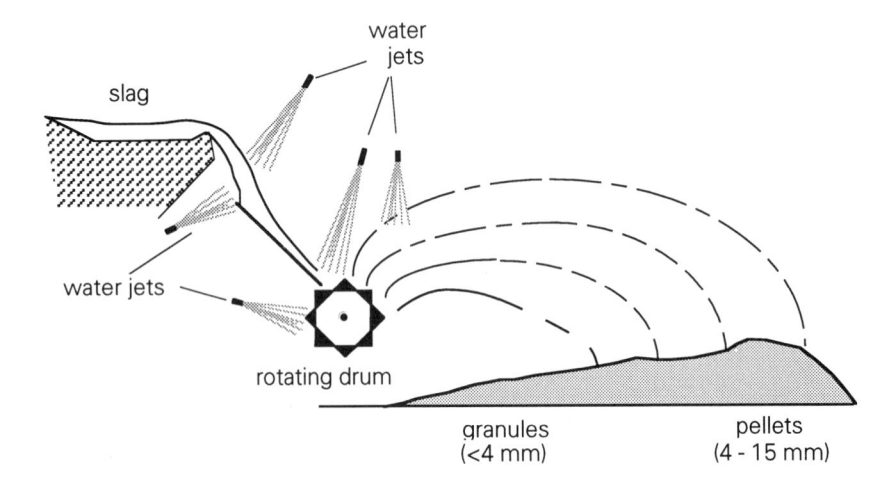

Figure 7.2 Pelletizer for the production of blast furnace slag products

Table 7.3 Chemical composition (wt. %) of some artificial pozzolanas (after Lea, 1970, pp.422 and 424)

	SiO_2	Al_2O_3	Fe_2O_3	CaO	$Na_2O + K_2O$	C
Fly ash	47	18	19	7	–	–
Rice Hull Ash	95	–	–	–	1	4

then added to the cement, or kept in a slurry and added wet in the concrete mixer. The latter, known as the Trief process saves fuel and minimizes the agglomeration of the particles that would take place during the drying of the slag. The use of a finely powdered slag increases its activity (since it is due to surface reactions) and this also increases the strength of the concrete in which the slag is used.

Some typical compositions of artificial pozzolanas are shown in Table 7.3.

7.3.3 Theories on the origin of the activity of the pozzolanas

There are many theories of the origin of the activity of the pozzolanas. These are described fully by Lea (1970, Ch. 14).

Pozzolanic activity is not simply related to the chemical composition since materials with the same composition exhibit widely differing activity. In the case of trass, the activity seems to be related to the presence of zeolites, and in the volcanic ashes it is apparently related to the presence of the glassy phase that has a high internal specific surface area. (The origin of the microporosity in the glasses is not known. It has been postulated that the pores are due to the exsolution of gases which occurred when the volcanic glass was rapidly cooled during an eruption. Other theories link it with the nature of the gases that were exsolved out of the glass droplets, but it could be that a gel-like structure is formed by the selective leaching of some of the more soluble constituents of the glass by weathering processes.) In the case of diatomaceous earth (which is almost pure amorphous silica) the porous nature of the diatoms results in there being a very large surface area to volume ratio. In burnt clays and shales, the activity is apparently associated with the production of amorphous m-kaolin by the removal of structural water from the crystalline clay layers. Fine grinding of the burnt clay is necessary to increase the surface area.

The effect of heat on pozzolanic activity of natural materials is variable. Santorian Earth loses most of its activity when heated to >750 °C whereas Italian pozzolanas show increased activity by a brief heating to 300–700 °C, but lose activity if held at 600 °C for an extended time (Lea, 1970, p. 418).

7.3.4 Setting reactions of pozzolanas

True pozzolanas will exhibit cementitious reactions only in the presence of added lime. The setting reaction takes place by the partial dissolution of the pozzolana in water to produce silica in solution. The dissolved silica then reacts with the calcium hydroxide (**CH**) to form a **CSH** gel. This surface solution of the silica in water could be the reason why a high surface area is apparently necessary for pozzolanic activity. The lime can either be mixed with the pozzolana, as in the case of additions of pozzolanas to lime mortars, or produced by other reactions, as in the case of additions to pozzolanas to Portland cement based mixes. In the latter case it is the hydration reaction of **C_3S** to form lime that activates the pozzolanas.

The high calcium fly ashes and water cooled (glassy) slag will react with water to produce **CSH** in the absence of added lime, but the rate of reaction is far too slow to be of any commercial use as a building material. The hydration reactions of the slag can be accelerated by solutions of alkalies or sulphates, thus the alkaline conditions that are present due to the hydration of Portland cement, and the presence of gypsum (calcium sulphate) will activate the hydraulic potential of the blast furnace slag.

7.3.5 Pozzolana additions to mortars

Lime-pozzolana mortars are produced by adding water to a mixture of lime, sand and a pozzolana. The reaction between the pozzolana and lime to produce the **CSH** gel produces a hydraulic setting mortar. The setting times of the lime-pozzolana mortars are variable. The initial set commences between one and three hours after mixing, and the final set occurs 10 to 12 hours after mixing. In the absence of the pozzolana, the mortar will harden very slowly by reaction with carbon dioxide.

In Portland cement-pozzolana mortars, the lime necessary for the setting reaction comes mainly from the hydration of the **C_3S** in the Portland cement.

7.3.6 Pozzolana additions to concrete

Partial substitution of Portland cement in concrete by pozzolanas results in an improvement of the workability of the concrete which means that a lower water/cement ratio can be used.

The rate of development of strength is affected because the pozzolana hydrates at a slower rate than the Portland cement. In the case of blast furnace slag this can be partially counteracted by grinding the slag to a finer state. Figure 7.3 shows that there is an initial reduction in strength, but, with time, the strength of the mix exceeds that of the cement without any pozzolana.

ADDITIONS OF POZZOLANAS

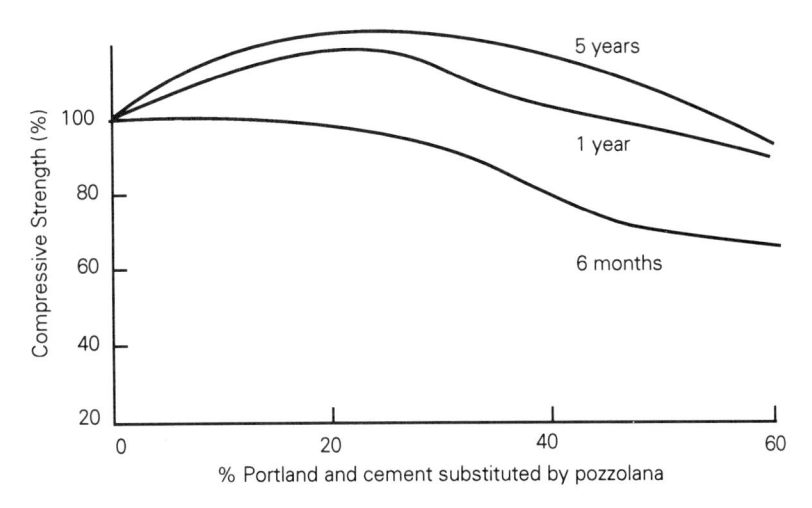

Figure 7.3 The effect of pozzolana substitution for Portland cement on the strength of concrete stored in water at 18 C (redrawn from Lea, 1970, p. 436)

The pozzolanic reaction between fly ash and the calcium hydroxide produced by the hydration of the cement commences after about 14 days and, depending on the activity of the fly ash, extends to about 150 days, after which time the fly ash particles are almost completely disintegrated by the chemical reaction with the calcium hydroxide to form **CSH**.

The slower hydration rate of the pozzolana which is substituted for some of the cement leads to a reduction in the rate of evolution of heat as shown in Table 7.4. This is useful in large constructions, such as dams, where heat evolution is always a problem and cooling systems have to be installed.

The amount of slag substituted for Portland cement is normally between 25% and 70% by mass, but slag contents of 50% to 90% may be used for the manufacture of a low-heat Portland-blast furnace cement. Similarly 10–20% of fly ash (by volume) is used to replace Portland cement in structural concrete, whereas >50% replacement can be made in mass concrete in cases where early strength is not required (Berry et al., 1994).

Table 7.4 Heat of hydration of blast furnace stage/Portland cement mixtures (after Reeves, 1985, p.65)

Total heat (kJ/kg) liberated after	0	30	% slag, by mass 40	50	70	90
7 days	360	325	305	255	200	125
28 days	400	395	360	325	260	185

SOME ADDITIVES (ADMIXTURES) USED IN MORTAR

Table 7.5 Time for 0.1% expansion of the cement by two forms of sulphate attack (Lea, 1970, p. 440)

Cement		Time for 0.1% expansion (weeks)	
Portland cement (%)	Pozzolana (%)	5% Na_2SO_4 solution	5% $MgSO_4$ solution
100	0	18	10
80	20 (burnt clay)	52	20
60	40 (burnt clay)	>200	220
80	20 (trass)	50	14
60	40 (trass)	>200	170

Another reason for the use of pozzolana-Portland cement concretes in such massive constructions is that of economy as the pozzolana is less expensive that the cement it replaces.

The resistance of the concrete to attack by sulphates and sea water is improved by the use of pozzolanas. The exact reasons for the increased resistance is uncertain, but it is thought to be due to the reduced content of free lime in the Portland cement/pozzolana mixture which would normally be involved in the sulphate attack. During the pozzolana/lime reaction an impervious layer is formed around the lime crystals, and this layer reduces their availability for sulphate attack reactions.

The effect of partial substitution of Portland cement by pozzolanic materials on attack by sulphate solutions measured by the time taken for 1% expansion is shown in Table 7.5.

The chemical resistance of slag/Portland cements towards attack by sulphates is greater than that of Portland cement. However the ratio of slag/Portland cement is important; there is little increase in resistance up to 50% slag content, but it increases rapidly when this amount is exceeded. The increased sulphate resistance also means that it can be used in sea-water environments (Neville and Brooks, 1987, p. 29).

7.4 SUPERPLASTICIZERS

Concretes with high workability are needed when they are to be used to cast slabs for floors etc., which often contain congested steel reinforcement. In this case, the concrete is pumped into place. In order to obtain concrete of sufficient strength, a concrete with a slump of between 50 and 75 mm has to be used, and to force the concrete to flow in-between the reinforcement, vibration of the concrete is necessary. The workability of the concrete could be increased by increasing the water/cement ratio, but this results in a loss of strength and increased segregation of aggregate.

Plasticizers (water-reducers) can be added to the concrete to overcome this problem. Addition of these materials results either in a lowering of the water/cement ratio without a simultaneous reduction in workability, or an increase in workability for the same water/cement ratio. Plasticizers are soluble macromolecules such as carbohydrates, hydrocarbolic acids, sulphonated lignin, sulphonated naphthalene or melamine formaldehyde. The size of these molecules is several hundred times that of a water molecule and, when adsorbed onto the surface of the cement particles result in a decrease in the friction between the solid particles (Guo, 1994). However, although the workability can be improved by the addition of plasticizers these also act to retard the set and this can result in segregation of the aggregate.

These problems have now been overcome with the use of 'superplasticizers' or high-range water reducers. These are very high molecular weight polymers and their size is 10 000 to 100 000 times that of a water molecule. When these compounds are added to the concrete mix, they are adsorbed onto the surface of the hydrating cement particles. This adsorption breaks down the agglomerates of cement particles and causes them to become mutually repulsive (i.e. a colloid is formed). The adsorption and dispersion of the cement agglomerates releases the water that is normally held inoperative within the agglomerates and frees it for the hydration of the cement grains. The adsorbed sheath of polymer around each particle not only reduces the friction between the particles but also has a 'smoothing' effect on the surface of the grains (Guo, 1994) allowing them to move more easily and forming a better aggregate/cement mixture, as shown in Figure 7.4.

Mindess (1994) lists three major types of superplasticizers, namely lignosulphonate based, polycondensates of formaldehyde and melamine

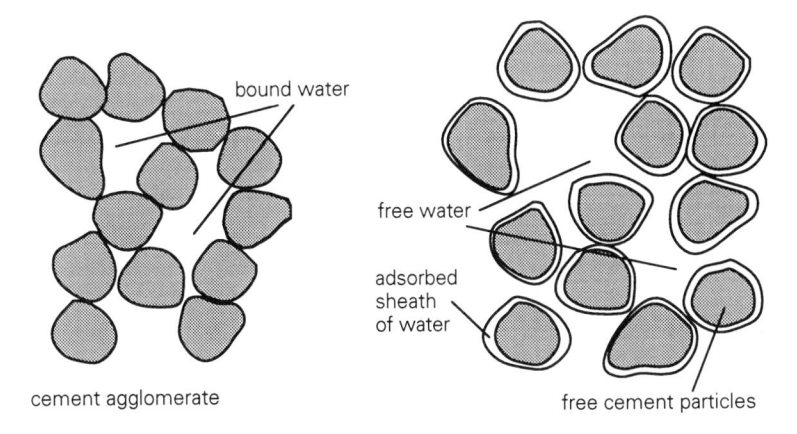

Figure 7.4 Effect of superplasticizers in breaking up cement agglomerates

SOME ADDITIVES (ADMIXTURES) USED IN MORTAR

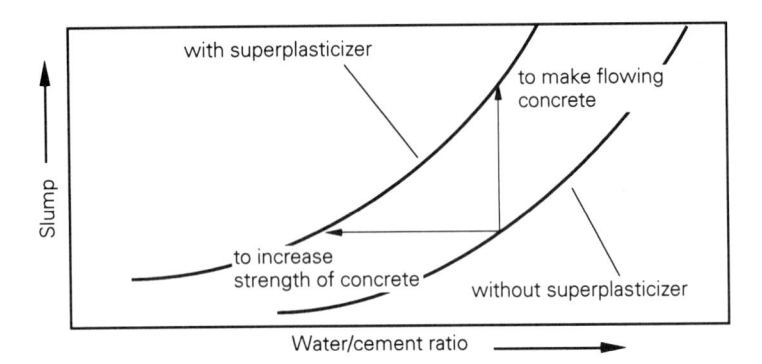

Figure 7.5 Use of superplasticizers to produce flowing concrete at constant water/cement ratio, or to increase the strength of concrete by a reduction in the water/cement ratio for constant slump

sulphonate and polycondensates of formaldehyde and naphthalene sulphonate. (The lignosulphonate based materials are used in conjunction with either of the other two types as they are not sufficiently efficient on their own.)

Addition of these agents to the same mix of cement, aggregate, sand and water increase the initial slump of 50–75 mm to >200 mm. The effect of the agents gradually wears off over a period of between 90 to 120 minutes, after which time the concrete reverts to its normal slump. The superplasticizers have no adverse effects on the hydration reactions of the cement. Tests carried out in Japan have shown that the strength of 'flowing concrete' after 11 years is increased by 54% compared with that of normal concrete. There is also no adverse effect on the drying shrinkage. ('Flowing concrete' is defined by ASTM as 'concrete that is characterized by a slump greater than 190 mm while maintaining a cohesive nature'.)

There are two types of admixtures described in the standard ASTM 1017. Type 1 produced flowing concrete with a normal setting time, whereas Type 2 is used in concrete where a retarded set is required (ACI Committee 212.4R, 1993).

The influence of superplasticizers on the slump of the concrete vs water content and on the development of early strength of the concrete are shown in Figure 7.5 (redrawn from Neville and Brooks, 1987, p. 158).

The use of this concrete in casting slabs has many advantages which include the removal of the need for vibration to compact and force the concrete between reinforcement. The mix behaves as a thixotropic non-Newtonian fluid, which means that it flows easily when a force is applied, but ceases to flow when the force is removed. These flowing concretes are more easily pumped and greater volumes can by placed in any given time. Another advantage is that the lower water/cement ratio, achieved

Table 7.6 The effect of the addition of superplasticizer to concrete mix ratios

	Mix A (with superplasticizer)	Mix B (without superplasticizer)
Wt. cement/m^3	350 kg	410 kg
Wt. sand/m^3	655 kg	530 kg
Wt. coarse/m^3	1190 kg	1190 kg
Water/m^3	185 litre	215 litre
Slump	200 mm	200 mm
Agg./cement ratio	5.27	4.19

with the use of superplasticizers, results in an increase of the density of the concrete and this, in turn, means that the concrete will have increased resistance to corrosion and increased durability.

The main disadvantage of flowing concrete is that it is not suitable for concreting on slopes >3°. There is also a need for the formwork to be placed more carefully, since the concrete will flow through any small gaps in the formwork.

The effect of the admixture of a superplasticizer on the concrete mix ratio can be shown by a comparison of the types of mix required to give the same slump and specified strength, as shown in Table 7.6.

It can be seen that the conventional mix (B) requires an extra 60 kg of cement and 30 litres of water to maintain the specified strength at a workability of 200 mm. The addition of the superplasticizer to mix A has increased the slump from 50–75 mm (which the mix would have had without the superplasticizer) to 200 mm.

The use of superplasticizers also tends to reduce the drying shrinkage of the concrete. The shrinkage of the concrete after setting is a function of the quantity of water added as well as the water/cement ratio. The lower the water content, and the higher the aggregate/cement ratio, the less the shrinkage that will occur on drying.

One of the major problems in the use of superplasticizers to improve the workability of the low water/cement mixes is that the cement and the plasticizer have to be chosen to be compatible. If this is not done, the effect of the superplasticizer wears off too soon and the workability of the mix can be lost before the concrete is placed. At present the only reliable way to solve this problem is to carry out tests on trial concrete mixes (Aïtcin and Neville, 1993).

SOME ADDITIVES (ADMIXTURES) USED IN MORTAR

QUESTIONS

1. What are accelerants and retardants?
2. What are the effects of calcium chloride additions to concrete?
3. Would you specify the use of calcium chloride in mixes of concrete to be used in reinforced internal walls and floors in a building? Give reasons for your answers.
4. What is the difference between 'entrained' air and 'accidental' air in a concrete mix?
5. How does entrained air increase the frost resistance of concrete?
6. What is a pozzolana?
7. What is the difference in the hydraulic setting reactions of Portland cement and a pozzolanic material?
8. What features do pozzolanic materials have in common?
9. Describe how pozzolana/lime mortars and pozzolana/Portland cement mortars set.
10. Why does the substitution of some of the Portland cement in a concrete mix by a pozzolanic material lead to an initial reduction in strength, but at later times may increase the strength of concrete to higher values than those attained without any substitution?
11. What is the essential difference between the blast furnace slag used as an aggregate in concrete and that used for partial substitution of cement in a concrete mix?
12. What are plasticizers and superplasticizers, and what effects do they have on fresh concrete?
13. What is 'flowing' concrete, how is it made and what can it be used for?
14. How is the consistency of flowing concrete measured?

References

ACI Committee 212 (1989) Chemical admixtures for concrete. *ACI Materials Journal*, **86**, May-June, 297–327.

ACI Committee 212.4R (1993) Guide for the use of high-range water-reducing admixtures (superplasticizers) in concrete. *Concrete International*, **15** (4), 40–7.

ACI Committee 232 (1994) Proposed report: use of natural pozzolans in concrete. *ACI Materials Journal*, **91** (4), 418–26.

Aïtcin, P.-C. and Baalbaki, M. (1995) Concrete admixtures - key to components of modern concrete. In A. Aguado, R. Gettu and S. P. Shah (eds), *Concrete Technology: New Trends and Industrial Applications*, E. & F. N. Spon, London, pp. 33–47.

Aïtcin, P.-C. and Neville, A. (1993) High-performance concrete demystified. *Concrete International*, **15** (1), 21–6.

Alonso, J. L. and Wesche, K. (1991) Characterization of fly ash. In K. Wesche (ed.), *Fly Ash in Concrete, Properties and Performance*, E. & F. N. Spon, London, pp. 2–23.

REFERENCES

Attiogbe, E. K. and Farzam, H. (1994) Extended set control of concrete. In P. W. Brown (ed.), *Cement Manufacture and Use*, American Society of Civil Engineers, New York, pp. 137–46.

Berry, E. E., Hemmings, R. T., Zhang, M.-H. *et al*, (1994) Hydration in high-volume fly ash concrete binders. *ACI Materials Journal*, **91** (4), 382–9.

Guo, C.-J. (1994) Early age behaviour of Portland cement pastes. *ACI Materials Journal*, **91** (1), 13–25.

Lea, F. M. (1970) *The Chemistry of Cement and Concrete*, 3rd edn, Arnold, London.

Malhotra, V. M. (1993) Fly ash, slag, silica fume and rice-husk ash in concrete: a review. *Concrete International*, **15** (4), 23–8.

Mindess, S. (1994) Materials selection, proportioning and quality control. In S. P. Shah and S. H. Ahmad (eds), *High Performance Concretes and Applications*, Edward Arnold, London, pp. 1–23.

Neville, A. M. and Brooks, J. J. (1987) *Concrete Technology*, Revised, Longman Scientific & Technical, Harlow, Essex, UK.

Reeves, C. M. (1985) The use of ground granulated blast furnace slag to produce durable concrete. In *Improvement of Concrete Durability*, Telford, London, p. 61.

Regourd, M. (1983) Slags and slag cements, *Journal of Materials Education*, **5**, 688–713.

Williams, J. T. and Swail, R. (1988) Air entraining admixtures for concrete. In P. C. Hewlett (ed.), *Cement Admixtures*, 2nd edn, Longmans Scientific & Technical, Harlow, Essex, UK, pp. 28–55.

8 High-performance concrete

In recent years there has been a great deal of research and development into the methods of production of high-strength mortar and concrete. The term 'high-strength' needs to be defined, since its value will depend on what strength was considered 'normal' at any period of time. For about 100 years (up to the 1970s), concrete with a 28-day compressive strength of 20–30 MPa was produced on a routine basis. During this time, 'high-strength' concrete with strengths above 40 MPa could be obtained, but later this term was applied to concrete with strengths of 50–60 MPa. At that time these high-strength concretes were only produced when necessary and not on a routine basis. However, from the late 1970s, there has been a dramatic change and concretes of strengths of 90–110 MPa have been used in the construction of high rise buildings and bridges. Concrete of these strengths are now made and routinely used, and occasionally 120 MPa concrete has been utilized. Since the term 'high-strength concrete' has had many different meanings over the years, it is now proposed that high-performance concrete (HPC) should be used to describe the high-strength materials that are now being produced (Aïtcin and Neville,1993).

The situation is now becoming more complicated as new attempts have been made to define what is meant by HPC which go beyond a simple definition referring to compressive strength only. Papworth and Ratcliffe (1994) define HPC as 'concrete that meets the requirements or goes beyond the limit of the normal performance range'. (The term 'limit' could apply to strength, density, durability, ductility, etc.) Mindess (1994) defines HPC in terms of strength and durability and quotes the definition of HPC for pavement (road) applications which includes strength characteristics, durability in freeze/thaw cycles and a maximum water/cementitious material ratio of 0.35. Forsh (1994) gives a broader definition;

> high performance concrete is a concrete made with appropriate materials combined according to a selected mix design and properly mixed, transported, placed, consolidated, and cured so that the resulting concrete will give excellent performance in the structure in

which it will be placed, in the environment to which it will be exposed, and with loads to which it will be subjected for its design life.

The latter definition takes into account the fact that all of the phases present in the HPC (cement paste, aggregate and the interfacial zones between the paste and aggregate) are pushed to their limits, so that all of the factors that can affect the properties of these phases must be taken into account (Mindess, 1994).

The need for further research into the properties of high-performance concrete is given in a technical report on high-performance construction materials and systems compiled for infrastructure in America in which it was found that a total of US $173.55 million was needed to fund the necessary research projects (Report 93–5011, 1993).

8.1 REDUCTION OF THE WATER/CEMENT RATIO

The strength of ordinary cement and concrete (normal strength concrete, NSC) is primarily controlled by the presence of flaws in the set material. These are present either because a high water/cement ratio had to be used to produce a workable mix, or because of air entrainment during the mixing of materials with lower water/cement ratios.

The use of a superplasticizer to reduce the water/cement ratio of the concrete to 0.2–0.3, while still retaining workability, is the basis of the production of HPC. In the previous chapter it was shown that the addition of a superplasticizer could be used to produce a flowing concrete with a slump of 200 mm from a mix that would otherwise have had a slump of 50–75 mm. Superplasticizers can also be added to a mix with low water/cement ratios (<0.3) which would normally be completely unworkable to increase their workability to that of conventional mixes.

The reduction in the water/cement ratio results in the production of a high-density cement matrix, and since there is no excess water, the capillary pores are largely eliminated in the paste and the compressive strength of the concrete can be increased to 60–80 MPa.

8.2 ADDITIONS OF MICROSILICA

In conventional concretes, the smallest size particles in the mix are the cement grains which range in size from 1 to 100 μm. A denser packing of the particles, and hence a higher strength, can be obtained if a still finer material is added to the mix to fill up the voids between the cement particles. Such a material is microsilica (silica fume) which is a by-product in the production of silicon metal and some ferrosilicon alloys. Microsilica consists of spherical amorphous silica particles with an average size of 0.1 μm. (Partial replacement of the cement by microsilica was originally done for economical reasons, since the microsilica, being a waste product, was much less expensive than cement. However it was soon noticed that these concretes had lower permeability, showed better frost resistance and had higher sulphate attack resistance and abrasion-erosion resistance than conventional mixes. The cost of microsilica varies but it is now more expensive than cement, and in some cases can be 10–15 times the cost of cement.)

The Norwegians were the first to test microsilica in concrete in the early 1950s, where it was used to line a tunnel in the Oslo alum shale region where the concrete was exposed to high sulphate, acid ground water (FIP Commission on Concrete, 1988). Since then, the use of microsilica has spread throughout the world, although there was some resistance to its use in Britain, primarily from the concrete suppliers who argued that the production of high grade mixes was not a commercial proposition (Ridout, 1990).

Various names are used for microsilica, these include condensed silica fume (CSF), ferrosilicon dust, arc furnace silica, silica flue dust, amorphous silica, volatilized silica and even very fine-grained siliceous fly ash (FIP Commission on Concrete, 1988).

8.2.1 Production of microsilica

The microsilica is a by-product of the submerged arc production of silicon and ferrosilicon alloys by the reduction of silica sand with a carbon source (e.g. charcoal from wood chips). The process is shown schematically in Figure 8.1.

This process is carried out open to the atmosphere and gaseous silica is formed by the oxidation of gaseous SiO which is produced in the vicinity of the electric arcs. The SiO_2 condenses from the vapour phase as a mist of fine droplets which, on solidification, form submicron spherical particles. In most cases the particles are glassy, but in silica produced from silicomanganese or calcium silicon over 25% of the material can be crystalline with the crystalline phases consisting of silicon carbide, calcite and quartz. The purity of the microsilica is dependent on the details of the process used and the product that is being made. In general the purer the silicon alloy that is being manufactured in the plant, the higher the silica content of the fume. In Australia, microsilica is obtained from two sources. The Tasmanian Electro Metallurgical Company (TEMCO) product typically contains 90% silica, 3% iron oxide, 2.5% carbon and 2.5% alumina. The other source, Microsilica Pty. Ltd (formerly Barrack Silica) in Western Australia (Burnett, 1991), does not give details of the purity; however as

ADDITIONS OF MICROSILICA

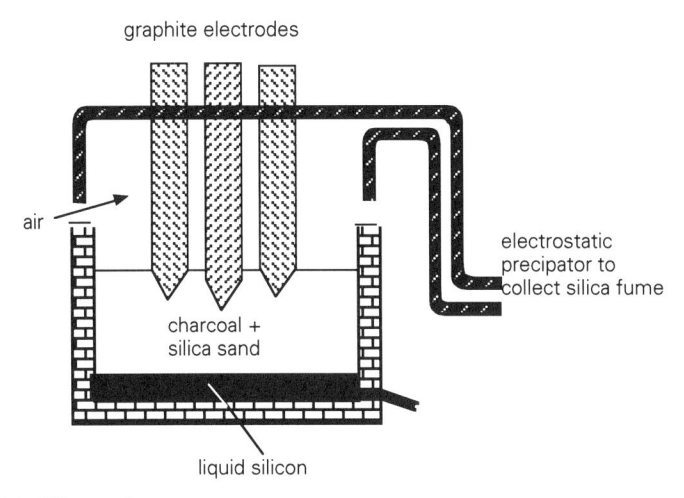

Figure 8.1 Silicon plant

the plant produces very high grade silicon by the use of Jarrah charcoal, it would be expected that the silica is of corresponding high purity.

The purity of microsilica obtained from various manufacturing processes is shown in Table 8.1.

The major impurities present in microsilica from silicon production are C, MgO, CaO and K_2O. In microsilica from ferrosilicon and ferrochromium silicon production they are Fe_2O_3, MgO, Al_2O_3 and C.

8.2.2 Effects of microsilica on the hardening and microstructure of concrete

In normal applications, part of the cement in a mix can be replaced by a smaller amount of microsilica without any loss of strength of the concrete. 1 kg of silica can replace 3–4 kg of cement, but its primary use is to

Table 8.1 Composition ranges of microsilica by product (Roberts *et al.*, 1993)

Product	wt% SiO_2 in microsilica by product
Silicon	91.5–98
Ferrosilicon (90% Si, 10% Fe)	89–96
Ferrosilicon (75% Si, 25% Fe)	70.8–94.9
Ferrosilicon (50% Si, 50% Fe)	83–91.1
Ferrochromium silicon	73–83

improve the properties of the concrete, both in the fresh and hardened state.

The microsilica has two overall effects on the concrete. Firstly it increases the density of packing, because it fills the spaces between the fine cement grains. Calculations have shown that for a 15 wt% replacement of cement by microsilica there are ~2 million particles of microsilica for each grain of Portland cement in a concrete mix (ACI Committee 234, 1995). Secondly, microsilica is an active pozzolanic material that reacts with the calcium hydroxide produced from the hydration of **C_3S** to form **CSH** gel.

Comparative measurements on the rate of hydration of the cement phases with and without microsilica have shown that the rate of hydration of the alite (**C_3S**) is increased if microsilica is present in the mix. It is thought that this is due to the slower formation of the gel membrane around the alite which results in rapid nucleation of **CSH** or calcium hydroxide (**CH**) on the microsilica which surrounds the alite particles. This is responsible for the rapid increase in early strength that the pastes exhibit. It is only at a later stage (several hours to days) that a pozzolanic reaction occurs between the microsilica and the **CH** to form more **CSH** gel. This results in a decrease of the permeability of the concrete (Roberts et al., 1993).

The presence of microsilica also affects the microstructure of the concrete. The most profound effect is observed at the interfaces between the various components in the concrete. In ordinary concrete, there is often a porous zone adjacent to the aggregate particles which contains a high concentration of orientated **CH** particles that originated from the nucleation of this phase at the aggregate interface. This zone is depleted of **CSH**, and hence is a zone of weakness. In microsilica concretes this zone is absent, since the microsilica acts as a nucleating agent for the **CH** so that it is dispersed throughout the paste. A similar nucleating effect has been observed when carbon black particles of the same size and shape to the microsilica were added to concrete (Goldman and Bentur, 1993). Microsilica also consumes the **CH** in the pozzolanic reaction and replaces it with **CSH**. This effectively increases the bond between the aggregate and the paste and hence further increases the strength of the concrete due to additional densification which take place up to two months after pouring (Goldman and Bentur, 1993; Roberts et al., 1993). The improved aggregate/paste bond also results in the improvement of the resistance to abrasion of the concrete because the aggregate particles are not so easily pulled out of the matrix. It also results in the ultimate strength of the concrete now being dependent on the strength of the aggregate.

Microsilica can also be added to sprayed Shotcrete or Gunite mortar mixes in order to reduce the rebound in dry-mix, increase the thickness of the layer that can be applied in one pass, improve cohesion, increase the compressive and flexural strengths and also increase resistance to chemical attack (Wolsiefer and Morgan, 1993).

STRUCTURE OF HIGH-PERFORMANCE CONCRETE

8.3 PRODUCTION OF HIGH-PERFORMANCE CONCRETE

Due to the very high surface area of the microsilica, its addition to a concrete mix requires an increase in the amount of water that has to be added to the mix to obtain the required workability. This means that high-range water reducing agents (superplasticizers) have to be added in order to reduce the water demand so that the required high strength and durability can be achieved. The combination of microsilica and superplasticizer produces HPC.

Aïtcin and Neville (1993) believe that microsilica is not an essential ingredient in all HPC mixes since strengths of 60 to 80 MPa can be achieved with the use of superplasticizers alone. However, the use of microsilica does simplify the production of HPC of these strengths, and for higher strengths it is essential.

As with normal strength concrete, additions of other pozzolanic and cementitious materials, such as blast furnace slag and fly ash, can reduce the cost of the high-performance concrete and also can produce other benefits. The lower reactivity of these materials, when compared with Portland cement means that the rheological properties of the mix can be better controlled, and also the amount of expensive plasticizer that has to be added can be reduced (Aïtcin and Neville, 1993). The beneficial effect of additions of microsilica, in particular, and other pozzolanic materials on the strength of the transition zone between the surface of the aggregate and the cement paste is also a key factor in the production of HPC (Domone and Soutsos, 1994).

With the additions of blast furnace slag, fly ash and microsilica, all of which are cementitious materials, the term water/cement ratio needs to be modified to water/(cement + cementitious materials) ratio. It is now becoming common to quote both ratios when describing a concrete mix.

The choice of the aggregate is of great importance in HPC, because it is the strength of the aggregate that now limits the strength of the HPC (unlike normal strength concrete, in which the strength is controlled by the presence of pores in the mortar matrix and the strength of the cement/aggregate interface). This is demonstrated by examination of fracture surfaces of HPC which indicates that the cracks pass through the aggregate particles more often than around them (Burnett, 1991). The compressive strength of the mortar with the incorporation of silica fume and superplasticizer can be as high as 300 MPa.

8.4 STRUCTURE AND PROPERTIES OF HIGH-PERFORMANCE CONCRETE

The low amount of water that is added to the mix (water/cement ratio as low as 0.2) results in partial hydration of the cement grains, and in

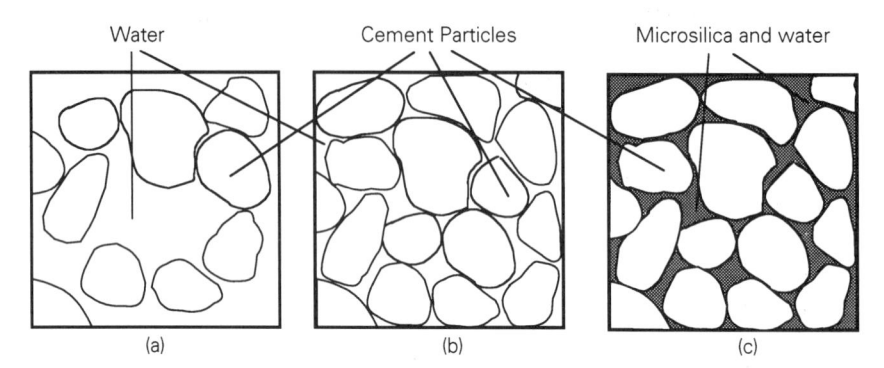

Figure 8.2 The structure of (a) cement paste, (b) superplasticized cement paste and (c) HPC paste systems

microsilica/cement mixes, much of the cement never fully hydrates. The hydration product that is formed is mainly **CSH** gel which appears microscopically homogeneous and contains a much lower Ca:Si ratio than the normal gel. Very few $Ca(OH)_2$ crystals are formed. No air voids exist in the mix and only very small capillary pores exist. It should be noted that the total porosity is not changed by microsilica additions, but there is a great reduction in the size of the capillary pores which, in turn, leads to a reduction of connected pores and an increase in isolated pores. This leads to a reduction in the permeability of the concrete to both liquids and vapours (ACI Committee 234, 1995). A relatively small addition of microsilica (5%) results in a 500% reduction in the permeability of concrete to water from 3×10^{-11} m/sec to 6×10^{-14} m/sec (Luther and Smith, 1991).

A model of the paste structure is shown in Figure 8.2.

The high compressive strength of the HPC is due to the low water/cement ratios that can be used and also the effect of the microsilica on the microstructure of the paste. The use of microsilica makes the production of 100 MPa concrete a routine matter. The contribution of the microsilica to the strength development of the concrete is almost complete 28 days after mixing (ACI Committee 234, 1995). The long-term stability of HPC has been a matter of debate, but tests carried out by Russell (1994) on 18-year-old HPC have shown no deterioration in properties.

The low permeability has the effect of increasing the durability of the concrete. The beneficial effect of this on the chemical resistance and the corrosion of reinforcement embedded in the concrete will be discussed in Chapters 12 and 13.

The major problem with the HPC is that when it fails under high loads the failure is catastrophic. This sudden failure can be overcome by the incorporation of fibres into the mix. These fibres do not contribute much to the strength of the cement, but do have a large effect on the toughness. The types and properties of fibres that are used in cements and concretes will be described in Chapter 11.

QUESTIONS

1. What is meant by 'High-performance Concrete' (HPC)?
2. How can superplasticizers be used to increase the strength of concrete?
3. Why is it essential to use a superplasticizer in conjunction with microsilica to produce high-performance concrete?
4. What is microsilica?
5. What effect does microsilica have on the early setting of concrete?
6. What effect does microsilica have on the aggregate/paste bond in concrete?
7. Why should the water/ (cement + cementitious materials) ratio be used instead of the water/cement ratio when describing HPC?
8. What causes the compressive strength of HPC to be higher than normal strength concrete?
9. What is the main factor that limits the strength of HPC?
10. If, in HPC, most of the cement never hydrates because of the low water/cement ratio used, why cannot the amount of cement added to the mix be reduced?

References

ACI Committee 234 (1995) Abstract of: Guide for the use of silica fume in concrete. *ACI Materials Journal,* **92** (4), 437–40.
Aïtcin, P.-C. and Neville, A. (1993) High-performance concrete demystified. *Concrete International,* **15** (1), 21–6.
Burnett, I. (1991) Silica fume concrete in Melbourne, Australia. *Concrete International,* **13** (8), 18–24.
Domone, P. L. J. and Soutsos, M. N. (1994) An approach to the proportioning of high-strength concrete mixes. *Concrete International,* **16** (10), 26–31.
FIP Commission on Concrete (1988) *Condensed Silica Fume in Concrete,* 37 pp, Thomas Telford, London.
Forsh, S. W. (1994) High performance concrete – stretching the paradigm. *Concrete International,* **16** (10), 33–4.
Goldman, A. and Bentur, A. (1993), Effects of pozzolanic and non-reactive microfillers on the transition zone in high strength concretes. In J. C. Maso (ed.), *Interfaces in Cementitious Composites,* E. & F. N. Spon, London, pp. 53–61.
Luther, M. D. and Smith P. A. (1991) Silica fume (microsilica) fundamentals for use in concrete. In P. W. Brown (ed.), *Cement Manufacture and Use,* American Society of Civil Engineers, New York, pp. 75–106.

Mindess, S. (1994) Materials selection, proportioning and quality control. In S. P. Shah and S. H. Ahmad (eds), *High Performance Concretes and Applications*, Edward Arnold, London, pp. 1–23.

Papworth, F. and Ratcliffe, R. (1994) High-performance concrete - the concrete future. *Concrete International,* **16** (10), 39–44.

Report 93–5011 (1993), *High-Performance Construction Materials and Systems: An Essential Program for America and Its Infrastructure*, American Society of Civil Engineers, New York.

Ridout, G. (1990) Whither high strength concrete? *Magazine of Concrete Research,* **42** (153), 191–2.

Roberts, L. R., Grace W. R., & Co. (1993) Microsilica in concrete, I. In J. Skalny (ed.), *Materials Science of Concrete*, The American Ceramic Society, pp. 163–80.

Russell, H. G. (1994) Long-term properties of high-strength concrete. *Concrete International,* **16** (4), 57–8.

Wolsiefer, J. Sr. and Morgan, D. R. (1993) Silica fume in shotcrete. *Concrete International,* **15** (4), 34–9.

Physical behaviour of concrete after pouring | 9

After the concrete is poured, hydration of the cement continues and, after some time, excess water is lost. During these stages, the physical behaviour of concrete can be roughly divided into two sections:

- changes that can take place before setting of the concrete has commenced, namely sedimentation, bleeding and plastic shrinkage and
- changes which take place after the concrete has started to set, namely thermal cracking, drying shrinkage, creep and fatigue.

9.1 BEFORE SETTING HAS COMMENCED

9.1.1 Sedimentation and bleeding

The differences in density of the components in the concrete mix can result in the segregation (or sedimentation) of these components in the concrete. A mortar, with a water/cement ratio of 0.6 will have a density of about half that of a typical aggregate. Sedimentation will result in there being a higher concentration of the aggregate at the bottom of a section, with the upper section being rich in mortar. There can also be separation of the coarser particles within the mortar itself, so that water is displaced upwards. This effect is known as 'bleeding'. The extent to which this occurs is dependent on the water/cement ratio in the concrete. Concretes with a large slump (which is due to a high water/cement ratio and not to the addition of plasticizing admixtures) are prone to bleeding. In some cases this occurs to such an extent that there is a layer of clear water on the surface of the fresh concrete. This is not always to be avoided, as it does result in a lowering of the water/cement ratio in the bulk concrete. This in turn results in an increase in strength of the bulk concrete and a decrease in the number of capillary pores (which are due to the presence

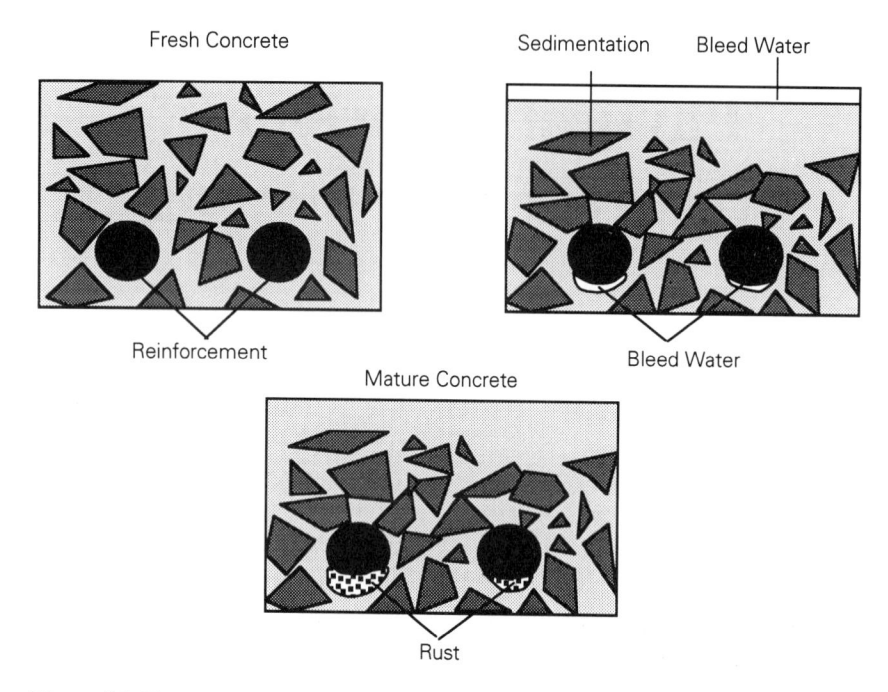

Figure 9.1 Entrapment of bleed water underneath reinforcement

of excess water). However there are some occasions where bleeding and sedimentation can have deleterious effects (Lees, 1992). These include:

(a) Corrosion of steel reinforcement in the set concrete

Segregation can lead to the entrapment of bleed water underneath steel reinforcement rods. This results in the formation of voids underneath the rods in the mature concrete, with the underside of the rods out of contact with the cement paste. This change of environment of the metal rods can result in corrosion (rusting) of the rods in this area, as shown in Figure 9.1.

(b) The production of a porous surface layer on the concrete

The increase in water content of the top layers of the concrete has the effect of increasing the water/cement ratio in this region, and if the water is not removed by evaporation, this layer will have different properties from the bulk of the concrete. This is particularly a problem with floor slabs, and is made worse by too-early trowelling to produce a flat surface. The more porous surface layer will then have a lowered resistance to wear and abrasion.

The above problems can be removed by decreasing the water/cement ratio, either by the addition of admixtures to reduce the water content needed for a given workability or by the addition of more cement. The settlement effects can also be reduced by revibrating the concrete an hour or so after pouring.

The addition of microsilica to the concrete mix causes a dramatic reduction of bleeding and separation of the components in the wet concrete. The reduction in bleeding is primarily due to the high surface area microsilica adsorbing water onto itself and leaving very little free water in the mix; secondarily it is due to the microsilica physically blocking the pores in the wet concrete (ACI Committee 234, 1995). The more cohesive mixture of the concrete also improves its in handling and transportation during pouring.

9.1.2 Plastic shrinkage

If the rate of evaporation of the bleed water is greater than the rate of bleeding, then 'plastic shrinkage' can occur. This shrinkage is due to the removal of water by evaporation from the surface of the concrete and can result in the formation of cracks. Plastic shrinkage mainly takes place when large flat slabs are poured in hot or windy weather and are not immediately covered to prevent loss of water. Cracking due to plastic shrinkage can also happen if the concrete is poured onto a porous substrate which can draw out the water from the wet concrete by capillary action (Lees, 1992).

The lack of bleeding observed in high-performance concrete means that this concrete is extremely vulnerable to plastic shrinkage and cracking. Protective measures to eliminate evaporation of water from the surface must be taken as soon as the concrete is placed (FIP Commission on Concrete, 1988).

9.2 AFTER SETTING HAS COMMENCED

9.2.1 Thermal cracking

The hydration of cement is an exothermic reaction and results in the evolution of heat while concrete is setting. The rate of heat evolution is dependent on the type of Portland cement that is used and the amount of cement in the concrete. For massive concrete structures it is necessary to use a low-heat cement together with a pozzolanic material to slow down the rate of evolution of heat and prevent excessive temperature increase in the bulk of the concrete. Care should be taken to avoid large temperature differences within the bulk of the concrete. Such temperature differences can cause thermal cracking if the cooler outer layers are put in tension by the warmer thermally expanded inner layers. Thermal cracking can be prevented by providing means for the removal of heat from the interior of the structure, for example by the use of cooling pipes set within the concrete or by thermally insulating the outer layers so that the whole structure is maintained at the same temperature.

If the concrete is at a spatially uniform temperature but one that is increasing with time, then this will also cause bulk thermal expansion of the whole mass of concrete. The amount of the expansion is dependent on the mix proportions and the type of aggregate. Providing the concrete is not constrained and is free to expand and contract, this should not result in cracking of the concrete. Thermal cracking of constrained concrete often occurs when the concrete is cooling from a higher temperature reached during setting, for example when walls are cast on hardened concrete slabs. The cooling contraction of the new concrete is constrained by the slab and cracking at the junction of the base of the wall and the slab can occur (Lees, 1992).

9.2.2 Drying shrinkage

Shrinkage is defined as the time-dependent strain due to moisture loss at constant temperature in the absence of an external load and takes place after the concrete has set (Hansen and Young, 1991). This shrinkage is due to loss of water from the concrete by evaporation. Water in hydrated cement or concrete can be loosely described as free (or excess) water located in capillary pores, physically adsorbed water on the surface of the **CSH** gels and chemically bonded water in the products of hydration of the cement.

The change in volume of drying cement paste is not equal to the volume of the water removed. The loss of free water from the pores, which takes place first, causes little or no shrinkage. But as drying proceeds, physically adsorbed water is removed, and the change of volume of the unrestrained cement paste is approximately equal to the loss of water one molecule thick from the surface of all the gel particles. Since the 'thickness' of a water molecule is about 1% of the gel particle size, a proportional change in dimensions of a pure cement paste on complete drying would be expected to be about 1% but values only up to 0.4% have been observed (Lea, 1970).

In concrete the shrinkage depends also on the aggregate/cement ratio and the water/cement ratio, as shown in Table 9.1.

The dependence of the shrinkage on the aggregate/cement ratio shown in the table would be expected, since it is dependent on the loss of physically adsorbed water from the gels formed from the hydrated cement and the larger the amount of cement, the larger the shrinkage. The increased

Table 9.1 The dependence of linear shrinkage on the water/cement ratio and the aggregate/cement ratio (Lea, 1970)

	Shrinkage after 6 months ($\times 10^{-6}$) water/cement ratio			
Aggregate/cement ratio	0.4	0.5	0.6	0.7
3	800	1200	–	–
5	400	600	750	850
7	200	300	400	500

shrinkage observed at higher water/cement ratios could be due to the increased amount of hydration of the cement particles due to the higher water/cement ratios six months after pouring.

The addition of superplasticizers to concrete increases the shrinkage. This is due to the effect of the additive in breaking up agglomerated cement particles which increases the efficiency of hydration of the particles. The measured shrinkage of these concretes has been found to be dependent on the type of superplasticizer used (Brooks, 1989), but on average the shrinkage increases by a factor of 1.2.

The evaluation of the effect of the additions of superplasticizers together with microsilica to produce HPC is complicated by the effects of the lower water/ (cement + microsilica) ratios that are used in these concretes. The use of these lower ratios decreases the shrinkage and this can offset the effects of the superplasticizer additions. In a review of published data on the effects of microsilica additions on concrete (FIP Commission on Concrete, 1988) it was concluded that the shrinkage of concrete is little influenced by the microsilica content, at least up to 10% by mass.

Other reports indicate that the presence of microsilica leads to higher autogenous shrinkage, otherwise known as self-desiccation or chemical shrinkage (Aguado and Gettu, 1993; Brooks and Hynes, 1993). In this process the water is trapped in the fine pore structure in the concrete and is retained to react with the cement constituents rather than being lost to the environment by evaporation. The overall result of this is a reduction in drying shrinkage. It has been found that the addition of 8% microsilica can lead to an estimated 25% mean reduction of drying shrinkage in concrete (Le Roy and De Larrard, 1993).

9.2.3 Creep

In discussions of deformation of concrete, it is normal to distinguish between creep as a deformation occurring under a constant sustained stress, and relaxation, which is a decrease in stress with time under constant deformation. In the past, creep in concrete has been variously

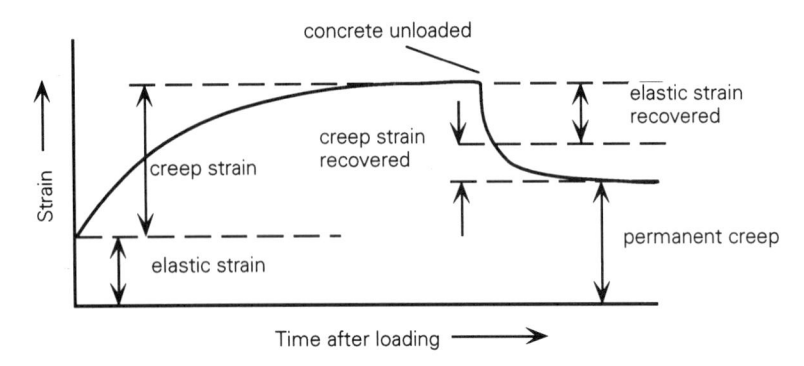

Figure 9.2 Creep of concrete

termed flow, plastic flow, plastic yield, plastic deformation, time yield and time deformation.

If concrete is placed under a sustained compressive load, then creep will occur. The creep occurs rapidly in the first few weeks of loading and then proceeds at a steadily decreasing rate. The creep is directly proportional to the applied stress up to about 20 MPa (approximately one-half of the ultimate compressive strength of normal strength concrete). The earlier that the concrete is loaded after setting, the greater is the creep obtained. A typical creep deformation after a year under load is approximately two to three times the elastic deformation that took place on the application of the load. One of the benefits of creep is that it leads to the reduction of internal stresses that can arise from drying shrinkage and this reduces drying cracking. However there are also problems, for example creep will result in a reduction of pre-stress in pre-stressed concrete members. Creep deformation must always be taken into account in the design of high-rise buildings, cantilever bridges with long spans or high piers and also in containment vessels.

Removal of the load will result in some creep recovery, as is shown in Figure 9.2.

Problems in the measurement of creep arise because, under normal circumstances, both creep and shrinkage due to drying occur simultaneously. It is found that the measured strain in such samples is always higher than that of the sum of the strain due to shrinkage and creep when they are measured separately. This is schematically shown in Figure 9.3.

The possible origins of this excess deformation are discussed fully by Neville (1981) and Wittmann (1982).

The creep of concrete under load is thought to be associated with the movement of physically adsorbed water on the gel particles (Marzouk, 1991). The major difference between shrinkage and creep is that shrinkage

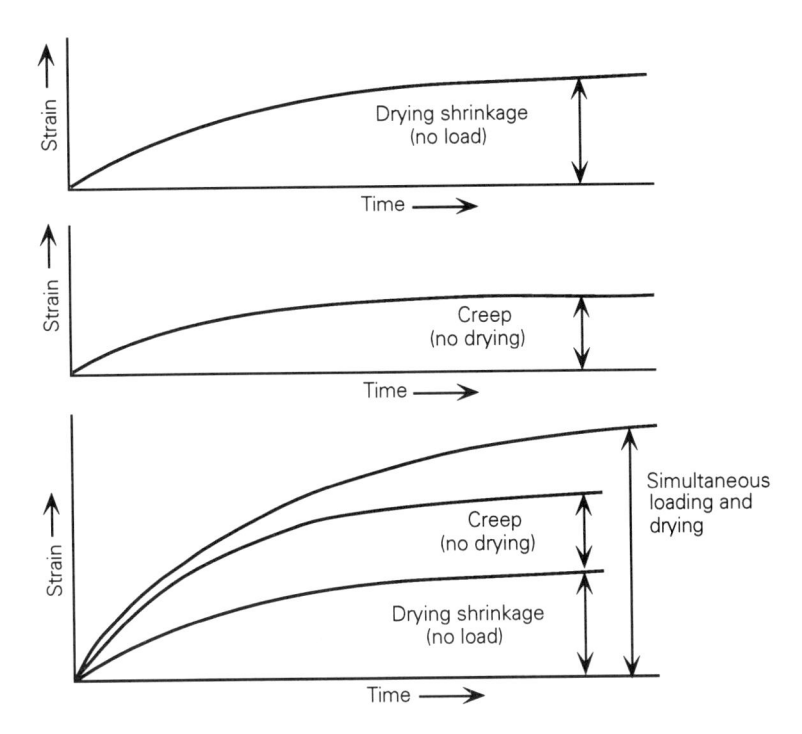

Figure 9.3 Simultaneous creep and shrinkage

is associated with the removal of the adsorbed water to the atmosphere, whereas creep is due to its movement within the concrete.

Since the origin of creep is similar to that of drying shrinkage it is influenced by similar factors, such as cement content, water/cement ratio, aggregate/cement ratio, etc. In general, for normal strength concrete, the amount of creep diminishes as the strength of the concrete mix increases.

The shrinkage (drying creep) of high-strength concrete to which microsilica additions are made is significantly reduced to almost zero due to the intense self-desiccation that takes place within the concrete, but the basic creep due to the movement of physically adsorbed water on the gel particles is similar to, or even greater than, that observed in normal strength concrete (Aguado and Gettu, 1993).

9.2.4 Fatigue

Fatigue of concrete can result in catastrophic failure taking place at less than the design load after the concrete has been exposed to a large number of stress cycles. These stress cycles can be brought about by wave action

Table 9.2 Factors that influence fatigue strength of concrete

Concrete composition	Effect of environment	Loading conditions	Mechanical properties
Air content	Temperature	Reversible	Compressive stress
Water/cement ratio	Moisture content	Variable stress	Tensile stress
Aggregate type	Aggressive agents	Constant stress	Elastic modulus
Cement content	Corrosion	Stress range	Modulus of fracture
Pozzolanas	Immersion	Load-wave form	Prestress
Curing conditions		Loading rate	Fibres
Age		Load amplitude	Precracking

on offshore platforms, vehicular traffic over bridges, machine vibrations, the landing of aeroplanes on runways etc.

The fatigue strength of concrete may be defined as a fraction of the static strength that the concrete can support for a given number of cycles (ACI Committee 215, 1993). The number of cycles to failure will increase as the concrete ages, this is due to the gain in static strength with time that occurs as the concrete continues to harden (Mays, 1992).

Each load cycle contributes to the damage of the concrete which occurs in stages. The formation of initial microcracks is observed at early stages of fatigue. Typically these cracks develop gradually and then stabilize to grow at a slow and steady rate until about 90% of the cycles to failure are reached, at which stage there is rapid crack growth and propagation until failure.

Failure of concrete by fatigue is characterized by much larger strains than observed in a comparable concrete under static loading. The fatigue strength of concrete for a life of 10 million cycles is approximately 55% of that under static loading (ACI Committee 215, 1993).

The factors that influence the fatigue strength of concrete are classified in Table 9.2 (from Mor *et al.*, 1992). There is no qualitative difference in the fatigue factors for normal and high-performance concrete.

QUESTIONS

1. What is sedimentation and bleeding of concrete?
2. Describe the advantages and disadvantages of sedimentation and bleeding of concrete.
3. How can sedimentation and bleeding be prevented or minimized?
4. What is the difference between drying shrinkage and plastic shrinkage?
5. Why is high-performance concrete (HPC) prone to plastic shrinkage?
6. What is the cause of thermal cracking and how can it be prevented?

7. Is drying shrinkage reversible?
8. What is creep of concrete?
9. What is the effect of the aggregate/cement ratio on the drying shrinkage and creep of concrete?
10. Would you expect HPC to exhibit more or less creep than normal strength concrete, and why?
11. Define the fatigue strength of concrete.
12. In some bridges the outer (slow) lanes overhang the supporting structure. Would you expect this to cause problems?

References

ACI Committee 215 (1993) Consideration for design of concrete structures subjected to fatigue loading – ACI 215R-74. In *ACI Manual of Concrete Practice Part 1 – 1993*, American Concrete Institute, Detroit, Mich., pp. 215R-1–215R-24.

Agaudo, A. and Gettu, R. (1993) Creep and shrinkage of high performance concretes. In Z. P. Bazant and I. Carol (eds), *Creep and Shrinkage of Concrete*, E. & F. N. Spon, London, pp. 481–92.

Brooks, J. J. (1989) Influence of mix proportions, plasticizers and superplasticizers on creep and drying shrinkage of concrete. *Magazine of Concrete Research*, **41** (148), 145–53.

Brooks, J. J. and Hynes, J. P. (1993) Creep and shrinkage of ultra high-strength silica fume concrete. In Z. P. Bazant and I. Carol (eds), *Creep and Shrinkage of Concrete*, E. & F. N. Spon, London, pp. 493–8.

FIP Commission on Concrete (1988) *Condensed Silica Fume in Concrete*, Thomas Telford, London.

Hansen, W. and Young, F. (1991) Creep mechanisms in concrete. In J. Skalny and S. Mindess (eds), *Materials Science of Concrete II*, American Ceramic Society, Westerville, OH., pp. 185–99.

Le Roy, R. and De Larrard, F. Creep and shrinkage of high-performance concrete: the LCPC experience. In Z. P. Bazant and I. Carol (eds), *Creep and Shrinkage of Concrete*, E. & F. N. Spon, London, pp. 481–92.

Lea F. M. (1970) *The Chemistry of Cement and Concrete*, 3rd edn, Edward Arnold, Ch. 10.

Lees, T. P. (1992) Deterioration mechanisms. In G. Mays (ed.), *Durability of Concrete Structures*, E. & F. N. Spon, London, pp. 10–36.

Marzouk, H. (1991) Creep of high-strength concrete and normal-strength concrete. *Magazine of Concrete Research*, **43** (155), 121–6.

Mays, G. C. (1992) The behaviour of concrete. In G. Mays (ed.), *Durability of Concrete Structures*, E. & F. N. Spon, London, pp. 3–9.

Mor, A., Gerwick, B. C. and Hester, W. T. (1992) Fatigue of high-strength reinforced concrete. *ACI Materials Journal*, **89**, 197–207.

Neville, A. M. (1981) Properties of concrete, 3rd edn, Pitman, London, Ch 6.

Wittmann, F. H. (1982) Creep and shrinkage mechanisms. In Z. P. Bažant and F. H. Wittmann (eds), *Creep and Shrinkage in Concrete Structures*, Wiley, Chichester (West Sussex), pp. 129–61.

10 Reinforced and prestressed concrete

10.1 REINFORCED CONCRETE

For design purposes, the tensile strength of concrete is assumed to be zero. In order to obtain appreciable tensile strength in a concrete structure, reinforcement bars (rebars) are embedded in the concrete. The rebars are normally made from mild steel (recently fibre-reinforced polymeric materials have been introduced). Tensile loads are supported by the rebars, as shown in the Figure 10.1.

For efficient load transfer from the concrete to the steel, the strength of the bond between the concrete and the rebar is important. The bond arises primarily from friction and adhesion between the concrete and the steel, but it may also be affected by the shrinkage of the concrete during setting. The mechanical properties of the rebar and its location in the concrete member are also important.

Generally the bond between steel and concrete is related to the quality of the concrete and the bond strength is approximately proportional to the compressive strength of the concrete. In order to obtain a better mechanical bond the rebars are normally ribbed (Figure 10.2). A thin layer of rust on the bars is also found to increase the bond between the

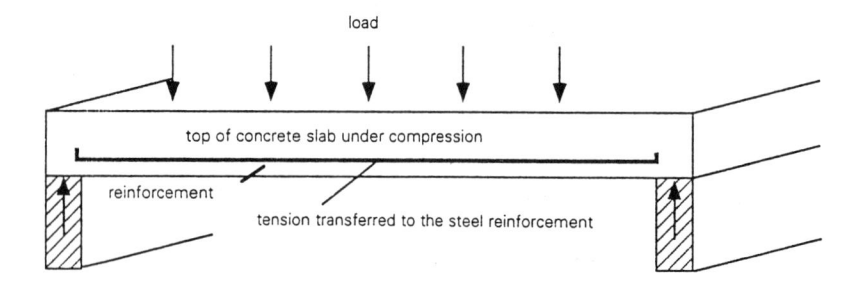

Figure 10.1 Transfer of tensile load to steel reinforcement

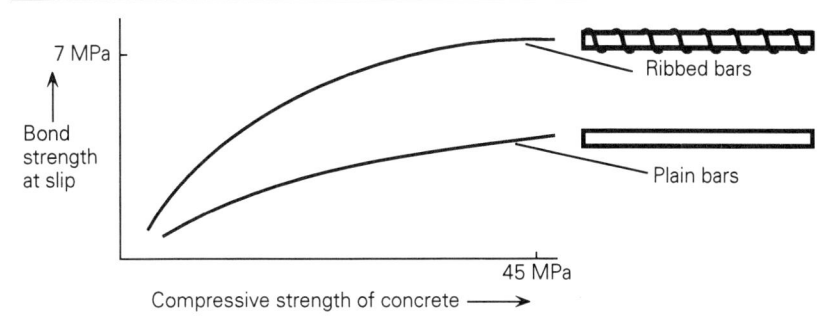

Figure 10.2 Effect of ribs on bond strength of concrete to reinforcement bars

concrete and the steel. This increase is probably due to the roughening of the surface of the steel by the rust formation.

If the steel rebar is surrounded with a sufficient coverage of good-quality concrete, the high pH of the concrete passivates the steel against corrosion. However continued corrosion of the rebar after placement sometimes takes place. This is normally due to either insufficient coverage of the bars with concrete, which could arise from movement of the rebars during the pouring of the concrete, or the use of a porous concrete which could be the result of using too high a water/cement ratio in the production of the concrete. Both of these causes permit the ingress of chloride ions, water and carbon dioxide which can contribute to corrosion. The corrosion of steel rebars is discussed in more detail in Chapter 12.

10.1.1 Reinforcement

A wide range of steels are used for reinforcement ranging from low carbon steels (carbon content <0.25 wt%), medium carbon steels (0.25–0.60 wt% C) to high carbon steel containing over 0.6 wt% C (Alekseev et al., 1993). The rate of corrosion of the steel is dependent on a large number of factors which affect its microstructure, such as chemical composition (content of carbon, silicon, manganese, copper etc.), thermal processing of the steel (hot or cold rolling, thermal hardening etc.) and welding during fabrication. Galvanizing and other protective treatments to prevent corrosion of rebars are sometimes used, but these often result in a reduction of the bond strength, since the good bond which occurs when the surface is rusty has been lost.

The use of fusion bonded epoxy-coated rebars (FBECR) is described by Reed and Atkins (1989). One of the arguments for the use of such coated rebars is that the poor quality of concrete used in some constructions has made such treatment necessary, but there is the problem that if the quality control of the concrete is poor, then there is also likely to be a similar lack of quality control in the use of the coated rebars. Case histories are given which illustrate this point. In the first case, a bridge in the Florida Keys, there were areas of insufficient concrete coverage as well high concentrations of chloride in the concrete and these were combined with the use of coated rebars which had been bent to such an extent during the installation, that cracking and disbonding of the coating had occurred. The final result was corrosion at these coating defects. The second case, which occurred in the Middle East, was the use of coated rebars that had been laying on site for six months prior to use. During this time, the coatings had degraded, and this led to corrosion of the reinforcement.

A detailed survey of the use of stainless steel (either as a cladding material over mild steel or as solid stainless steel rebars) instead of mild steel, epoxy-coated steel or even titanium, is presented by McDonald *et al.* (1995). They conclude that stainless steel bars exhibit excellent resistance to corrosion even in severe chloride environments. In the many studies that are reviewed in the paper, no cracking of concrete was reported due to corrosion of the stainless steel and there was no evidence of stress corrosion cracking of the steel itself. The increased cost in the use of stainless steel was considered to be reasonable and the use of such steel is warranted in conditions where there must be a guaranteed long-term resistance to corrosion.

Although steel is still the most popular material for use as reinforcement in concrete, there is much active research being carried out into the possibility of using fibre-reinforced plastics as rebars. It is intended that these be used in concrete that would be subjected to corrosive agents such as would occur in marine environments, or because of the use of deicing salts or in the environs of chemical plants (Chaallal and Benmokrane, 1992). The materials used range from hybrid fibre-reinforced plastic (FRP)/steel rebars (Nanni *et al.*, 1992), where the epoxy impregnated yarns of glass, aramid or carbon fibres are used to encase the steel wire fully, to glass-fibre-reinforced plastic (GFRP) rods made by a pultrusion process (Chaallal and Benmokrane, 1992) . Good adhesion to the concrete is achieved by bonding sand particles onto the surface of the rods. FRP fabricated into reinforcement grids have also been tested (Tao *et al.*, 1992; Larralde,1992; Schmeckpeper and Goodspeed, 1992). The articles refer to such advantages as corrosion resistance, light weight, high strength, good fatigue resistance etc. Erki and Rizkalla (1993b) list some of the disadvantages of the use of FRP for rebars as being the high cost (5 to 50 times that of steel), low modulus of elasticity and low failure strain. The low elastic modulus results in FRP-reinforced beams exhibiting much larger deflections than equivalent beams reinforced with steel rebars (Brown and Bartholomew, 1992). Pullout tests also indicate that the bond between the concrete and FRP is only about two-thirds that obtained between concrete and steel.

Continuous PAN-type carbon fibres (carbon fibres produced from polyacrylonitrile) which have been prefabricated into a three dimensional fabric impregnated with epoxy resin have been used by Zia *et al.* (1992) to reinforce concrete beams. The open weave fibre fabric reinforcement (including the epoxy coating) had a cross-sectional area of 4.19 mm^2 and was made up from 48 000 fibres. Concrete reinforced with this fabric exhibited the same ultimate strength and the same deflection at failure as a steel-reinforced beam, but in the post-cracking stage, the cracks were much smaller and more closely spaced in the carbon reinforced beams. The bond between the epoxy coated carbon fibre reinforcement and the concrete was stated to be excellent.

10.2 PRESTRESSED CONCRETE

In a normally loaded reinforced concrete beam, the reinforcement is designed to take the tensile forces in the bottom of the beam, which, together with the compressive force in the concrete at the top of the beam, resist bending of the beam under load. If the beam is overloaded, then cracks will appear in the concrete at the bottom of the beam (where the concrete is in tension) but the compressive load at the top of the beam may still be well below its failure value.

Prestressing puts the whole section into axial compression. This is done by embedding high yield point steel tendons under a tensile stress in concrete. The ends of the stressed tendons are securely anchored in the concrete and this results in the whole concrete section being put into compression. When the bending load is applied to the prestressed beam, the force that would cause tension and cracking in the bottom of a reinforced beam only reduces the compression which the prestress has already provided. The amount of compression at the top of the beam is increased, but this principle takes full advantage of the high potential compressive strength of the concrete. Its low tensile strength is of little consequence when prestressing is used.

The prestressing of concrete has the major advantage that lighter members can be used, this is particularly useful in bridges, long span roofs etc.

A prestressed member is generally designed to carry the working loads without the production of tension in the concrete. If the member is overloaded, then it has the capacity of almost complete recovery (provided that the yield point of the steel has not been exceeded) and tension cracks caused by the overload virtually disappear.

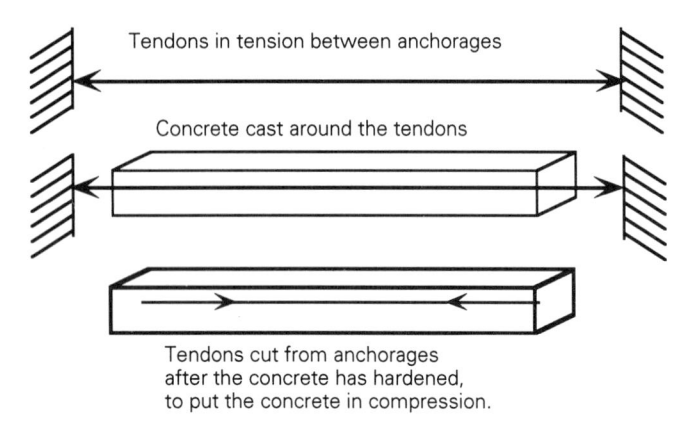

Figure 10.3 Schematic representation of pre-tensioning of concrete

10.2.1 Methods of applying the prestress

(a) Pre-tensioning

Pre-tensioning of a concrete beam is shown schematically in Figure 10.3.

The steel tendons are put in tension and the concrete cast around the steel. The tension is maintained until the concrete has attained sufficient bulk strength to resist the compression about to be imposed on it and sufficient bond strength to be able to transfer the prestress from the steel via the cement/steel bond. Once the concrete has attained the required strength, the wires are cut from their anchorages and the resultant tensile force in the tendons puts the concrete into compression. The ends of the steel must be then covered with a sufficient depth of concrete or mortar to prevent corrosion of the stressed steel.

Pre-tensioning is normally used only for factory production of precast units, it is seldom applicable to structures cast on site because of the time required for the concrete to attain sufficient strength, during this time the tendons have to remain attached to the tensioning devices.

(b) Post-tensioning

In the post-tensioning method, as shown in Figure 10.4, the prestressing tendon is either cast in the concrete, but initially prevented from bonding to the concrete by some form of sheath, or is threaded through holes already cast in the concrete.

The tendons (cables) are put into tension only after the concrete has hardened. One end of the cable is anchored and the stress is applied to the other using hydraulic jacks. Once the full prestress has been applied,

PRESTRESSED CONCRETE

Hole cast in concrete.

Post-tensioning tendon threaded through hole in set concrete, and tension applied.

anchoring grip

Grout pumped around strained tendon.

Tendon anchored by grip at other end and released

Concrete in compression

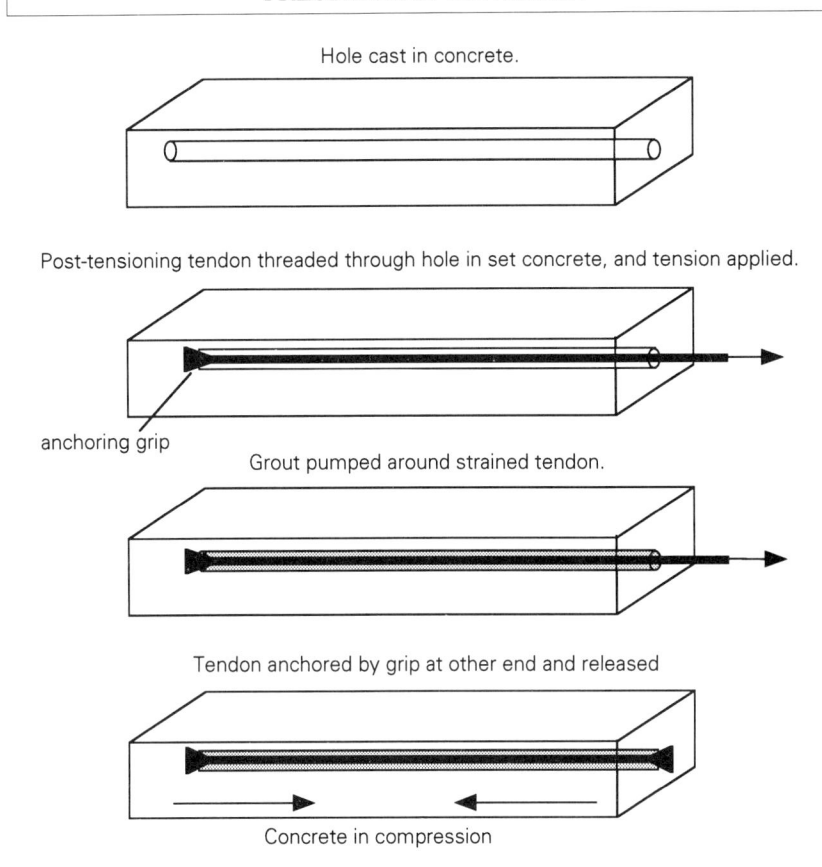

Figure 10.4 Schematic representation of post-tensioning of concrete

grouting is forced through the holes which then bonds the whole cable to the concrete. The end of the stressed tendon is then clamped with grips which maintain tension by a wedging action.

The advantage of post-tensioning is that the concrete has already hardened, and some of the drying shrinkage has already taken place, however, there will still be some losses of stress due to creep of the concrete. It is a rapid procedure, when compared with pre-tensioning and therefore suited to structures cast on site.

Post-tensioned cables are usually grouted up immediately after straining and anchoring. The grout consists of neat cement and water. It is forced in one end until the grout issues from the other. The ends are then sealed to ensure that none of the grout flows out. In theory this should result in the duct being completely filled with grout, however examination of post-tensioned bridges has revealed that this is not always the case (Mallett,

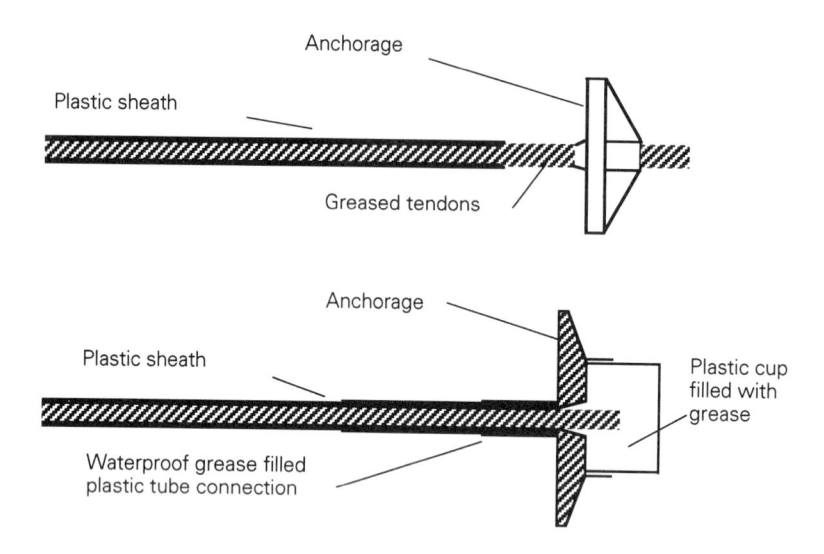

Figure 10.5 Unbonded tendons

1994). There was still sufficient grout present to prevent corrosion of the tendon, but it presented a potential problem with corrosion if ingress chlorides could take place (for example on post tensioned concrete bridges where de-icing salts are used).

It is essential that the ends of the structure, where the anchorages are fixed, have sufficient coverage of concrete or mortar to prevent local corrosion.

In the past 30 years there has been a trend to use ungrouted (unbonded) tendons in post-tensioned structures. In the unbonded tendons, the steel is free to move relative to the structure. The mechanical end-anchorages serve to transmit the pre-stressing force to the structure. This has the advantage that the tendons can be monitored for stress and corrosion and can be replaced if necessary. This technique has been used in the construction of nuclear power plants in the USA (Schupack, 1991).

The original unbonded tendons were covered in a corrosion resistant grease and then wrapped in paper, but since the 1970s, plastic sheaths have been used as shown in Figure 10.5. The plastic sheathing may be extruded or made from heat sealed strip.

Care has to be taken in the installation of the tendons in order to avoid corrosion, since corrosion protection comes only from the presence of the corrosion inhibiting grease and not, as is the case of embedded tendons, from the high pH of the concrete or grout.

10.2.2 Tendons

A high tensile steel must be used for the tendons since they are loaded to just below the yield point, which should be as high as possible. The tendons may be hard-drawn wire, strand, or alloy steel bars similar in size to normal reinforcement. In this case it is essential that the prestressing steel and anchorages be free of mill scale, rust, oil etc. as these could result in stress corrosion cracking of the tendons under tension while in use in the concrete structure.

Recently fibre-reinforced plastic (FRP) has been used as reinforcing tendons (Erki and Rizkalla, 1993b; Sen *et al.*, 1994). Carbon, aramid and glass fibres impregnated with resin are formed into tendons. The advantages and disadvantages of using FRP for tendons are similar to those for rebars, but there is an added problem of the design of suitable anchorages for the tendons. FRP has a low compressive strength perpendicular to the fibres which makes anchorage without damage to the FRP difficult (Erki and Rizkalla, 1993a). Damage to the FRP at the anchorage can result in sudden brittle fracture, meaning that the prestressing operation can be very dangerous. Despite this, carbon fibre-reinforced plastic (CFRP) tendons have been used in the construction of a bridge in Calgary, Canada (Rizkalla and Tadros, 1994). The bridge was constructed in 1993 and fibre optic sensors, which were used to measure both temperature and strain were used to monitor the behaviour of the tendons during construction.

10.2.3 Quality of concrete

Maximum advantage is obtained from prestressing when it is used with high-strength concrete. Because of the dangers associated with the effects of corrosion of the steel tendons, calcium chloride additions should never be used to accelerate the setting of the concrete.

The properties of the concrete that result in its movement must be borne in mind when prestressing is to be used. The effects of the drying shrinkage, creep under load and the elastic deformation of the concrete at load transference will all tend to reduce the prestressing effect. The movements result in a shortening of the concrete which tends to relieve the prestressing effect of the wires or cables with time and it is necessary to increase the tension beyond that considered necessary for design purposes in order to counteract this gradual loss. The loss in prestress is greatest in fresh concrete but gradually diminishes with age. The limit of loss of prestressing due to the movement of the concrete is about 16%. The use of high-performance concrete with a low water/cement ratio will reduce the movements of the concrete due to these factors and therefore increase the efficacy of prestressing.

QUESTIONS

1. Why is reinforcement used in concrete?
2. Why are the mechanical properties of rebars and their location in a concrete beam important?
3. Why is a layer of rust desirable on the surface of rebars while prestressing tendons must be rust free?
4. Discuss the use of fibre-reinforced plastic (FRP) rebars in reinforced concrete.
5. Explain the difference between reinforced concrete and prestressed concrete.
6. Why is pre-tensioning normally carried out in a factory, whereas post-tensioning can be done on site?
7. Describe the effects of drying shrinkage and creep on pre- and post-tensioned concrete.
8. What advantages do unbonded tendons have over grouted tendons?
9. What are the advantages of using high-strength concrete in prestressed structure?

References

Alekseev, S. N., Ivanov, F. M., Modry, S. *et al.* (1993) *Durability of Reinforced Concrete in Aggressive Media*, A. A. Balkema, Rotterdam, pp. 218–26.

Brown, V. L. and Bartholomew, C. L. (1993) FRP reinforcing bars in reinforced concrete members, *ACI Materials Journal*, **90**, 34–9.

Chaallal, O. and Benmokrane, B. (1992) Glass-fibre reinforcing rod: characterization and application to concrete structures and grouted anchors. In T. D. White (ed.), *Materials Performance and Prevention of Deficiencies and Failures*, American Society of Civil Engineers, New York, pp. 606–17.

Erki, M. A. and Rizkalla, S. H. (1993a) Anchorage for FRP reinforcement, *Concrete International*, **15** (6), 54–9.

Erki, M. A. and Rizkalla, S. H. (1993b) FRP reinforcement for concrete structures, *Concrete International*, **15** (6), 48–53.

Larralde, J. (1992) Feasibility of FRP molded grating-concrete composites for one-way slab systems. In *Materials Performance and Prevention of Deficiencies and Failures*, American Society of Civil Engineers, New York, pp. 645–54.

Mallett, G. P. (1994) *Repair of Concrete Bridges*, Thomas Telford, New York, pp. 8–9.

McDonald, D. B., Sherman, M. R., Pfeifer, D. W. *et al.* (1995) Stainless steel reinforcing as corrosion protection, *Concrete International*, **17** (5), 65–70.

Nanni, A., Okamoto, T., Tanigaki, M. *et al.* (1992) Hybrid (FRP + steel) reinforcement for concrete structures. In T. D. White (ed.), *Materials Performance and Prevention of Deficiencies and Failures*, American Society of Civil Engineers, New York, pp. 655–65.

Reed, J. A. and Atkins, W. S. (1989) FBECR: The need for correct specification and quality control, *Concrete*, **23** (8), 23–7.

REFERENCES

Rizkalla, S. H. and Tadros, G. (1994) A smart highway bridge in Canada, *Concrete International*, **16** (6), 42–4.

Schmeckpeper, E. R. and Goodspeed, C. H. (1992) Splice/development length requirements for FRP grids used in the structural reinforcement of concrete. In T. D. White (ed.), *Materials Performance and Prevention of Deficiencies and Failures*, American Society of Civil Engineers, New York, 632–44.

Schupack, M. (1991) Corrosion protection for unbonded tendons, *Concrete International*, **13** (2), 51–7.

Sen, R., Spillett, K. and Shahawy, M. (1994) Fabrication of aramid and carbon fibre reinforced plastic pretensioned beams, *Concrete International*, **16** (6), 45–7.

Tao, S., Ehsani, M. R. and Saadatmanesh, H. (1992) Bond strength of straight GFRP rebars. In T. D. White (ed.), *Materials Performance and Prevention of Deficiencies and Failures*, American Society of Civil Engineers, New York, pp 598–605.

Zia, P., Ahmad, S. H., Garg, R. K. *et al.* (1992) Flexural and shear behaviour of concrete beams reinforced with 3-D continuous carbon fiber fabric. In H. W. Reinhart and A. E. Naaman (eds), *High Performance Cement Composites*, E. & F. N. Spon, London, pp. 495–506.

11 Fibre-reinforced cement and concrete

Fibres have been used to reinforce brittle building materials since ancient times. Straw was used to reinforce sunbaked bricks, horse hair to reinforce plaster, and more recently (AD~1900) asbestos fibres were used to reinforce Portland cement.

In western history the oldest recorded account of the use of fibre reinforcement is in the Old Testament of the Bible, Exodus **5**:6–7:

> And Pharaoh commanded the same day the taskmasters of the people, and their officers, saying, 'Ye shall no more give the people straw to make brick, as heretofore: let them go and gather straw for themselves.'

A chronological review of the earliest patents that have been taken out on fibre-reinforced cement and concrete is given by Naaman (1985). This includes a 1918 French patent by Alfsen in which is described the use of fibres (iron, wood etc.) to increase the tensile strength of concrete and it is suggested that the performance might be improved if the surface of the fibres were roughened or the ends of the fibres bent. In 1926 a patent was lodged by Martin who described the use of plain or crimped steel wires to strengthen concrete. In 1943, Constantinesco filed a patent in which he described the use of various kinds of steel fibre for increasing the toughness of concrete. He suggested applications for this concrete including air-raid shelters, army tanks and machinery foundations. The fibres he described are similar to the ones currently in use.

It was only after 1960 that the use of fibres in cement and concrete became widespread and the range of materials used to make the fibres increased. The fibres that are currently being used in cement and concrete can be broadly classified into two types (Swamy *et al.*, 1974):

- High modulus, high-strength fibres. These include fibres of steel, glass, asbestos and carbon which produce strong composites with cement,

impart strength and stiffness to the composites as well as increase the toughness.
- Low modulus, high elongation fibres. These include nylon, polypropylene and polyethylene. The composites produced with these fibres have large energy absorption characteristics. The fibres do not lead to any improvement in strength, but impart toughness and resistance to impact or explosive loading. (Cement and concrete containing these types of fibres should be strictly called fibre-toughened cement or concrete and not fibre-reinforced.)

Short, discontinuous fibres are normally used in fibre-reinforced concrete so the fibre-matrix bond is irregular and discontinuous. Apart from the fibre geometry other factors such as length/diameter ratio, fibre volume, orientation and fabrication techniques, profoundly influence the properties and the mode of failure of the fibrous composites. The role of the fibres is essentially to toughen the concrete by arresting any advancing cracks, thus delaying their propagation across the brittle matrix, and creating a distinct slow crack propagation stage. The ultimate cracking strain of the composite is thus increased to many times more than that of the unreinforced matrix.

The fundamental difference between the cement-based fibre composites and other fibre-reinforced materials, for example glass-fibre-reinforced polymers, is that in cement the matrix failure strain is only a fraction of the fibre yield strain. Also the matrix is porous, so the precise mechanism of how short discrete fibres reinforce the cement matrix has not yet been fully determined.

In the past, the main method used for the reinforcement of cement and concrete in a structure has been by the use of continuous reinforcing bars which are placed in the structure at locations such that they can withstand any tensile stresses. The incorporation of fibres as reinforcement is normally as randomly distributed fibres, and therefore they are not as effective as steel reinforcement in withstanding tensile stresses. But, since they are more closely spaced, they are much more efficient at controlling cracking of the cement matrix. Thus, conventional reinforcing bars are used to increase the load-bearing capacity of the concrete, and fibres are used primarily to increase the toughness by the control of cracking (Bentur and Mindess, 1990, p. 3).

These differences mean that there are applications in which fibre reinforcement can be better utilized than steel reinforcement bars. Some examples are as follows:

- In the manufacture of thin cement sheets. In this case the fibre content is normally in excess of 5% and the fibres act to increase both strength and toughness.
- In components that must withstand high-impact loading, for example,

blast proof shelters or precast piles that are sited by hammering them into the ground.
- In cases where cracking needs to be controlled, for example, aircraft runways, road surfaces etc.

Thus, in most cases, fibres are not used to increase the strength of a component (although an increase in the tensile strength may occur), but rather to increase its toughness (the energy absorption capacity of the material). As well as this, there may also be an increase in its impact resistance, fatigue properties and abrasion resistance. (As previously mentioned from this point of view the use of the term 'fibre reinforcement' is somewhat misleading, but its use is now widespread in the literature.)

11.1 TYPES OF FIBRE-REINFORCED CEMENT (MORTAR)

11.1.1 Ferrocement

Ferrocement may be considered as being in-between ordinary reinforcement and fibre reinforcement.

A fine steel mesh is used to reinforce the mortar. The mesh size of the steel can vary from 'chicken wire' (~20 mm) to 'fly wire' (~2 mm). The mesh is placed into a form having the final shape and the cement is applied by spreading or spraying a cement mortar paste onto the mesh. When the mortar has set another mesh may be fitted. It is possible in this way to build up several sheets of mesh reinforcement. It is of interest to note that a patent on the used of ferrocement was taken out in 1885 by Lambot in France, who used it for making rowing boats. Two of these boats still exist.

If properly designed and constructed, ferrocement sections compare very favourably with normal reinforced cements. If the ferrocement is placed under tension, numerous microcracks are formed, but these cracks are well under 100 μm and the ferrocement remains watertight and corrosion resistant under normal circumstances.

Ferrocement is used today for the manufacture of roof panels, boats etc.

11.1.2 Steel fibres

Steel fibres are used to reinforce cement, but their high cost compared with asbestos or glass restricts their widespread use. The fibres are usually made from cold-drawn wire.

The diameter of the fibres is typically from 50 to 250 μm and the lengths range from 3 to 50 mm. The quantity of fibres that can be added to the cement and still be uniformly distributed depends critically on the length/

TYPES OF FIBRE-REINFORCED CEMENT (MORTAR)

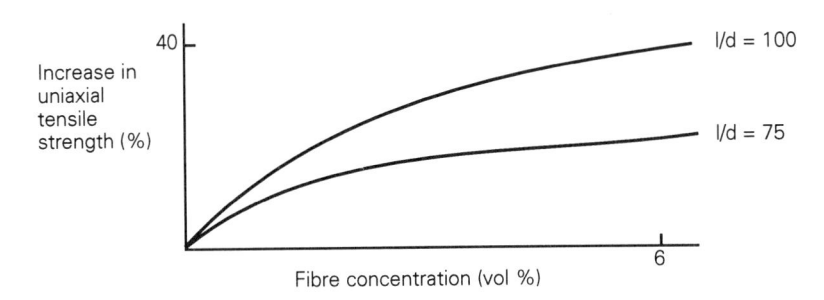

Figure 11.1 The effect of fibre concentration and l/d on the tensile strength of the composite

diameter ratio (l/d) of the fibres. For example, only 2 to 10 % (by volume) of fibres can be added to the cement paste for l/d between 100 and 200. This limitation is mainly due to the tangling of the long fibres which takes place during their mixing into the cement paste. Higher volumes can be added if the l/d is decreased, but this does not result in a large improvement in properties, because the shorter fibres have much reduced anchorage in the cement and fibre pull out occurs rather than tensile failure of the fibres (Krenchel, 1974).

The effect of the fibre concentration and l/d on the tensile strength is shown in Figure 11.1 while the effect of l/d ratio on the toughness is shown in Figure 11.2.

The interface between the steel fibres and the cement matrix has an important influence on the toughening of the cement by these fibres. It is known that if there is a weak interface between a fibre and the matrix, an approaching crack will be diverted along that interface to run parallel to the fibre. In the case of steel fibre-reinforced cement (SFRC), this weak interface is not the fibre/cement interface, but a porous layer formed close

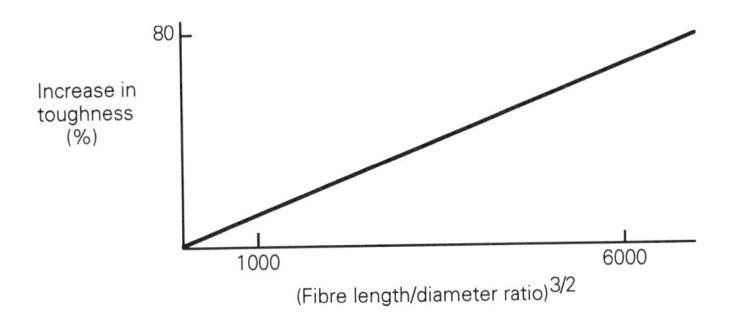

Figure 11.2 The effect of the l/d on the toughness of the composite (redrawn from Beaudoin, 1990, p. 95)

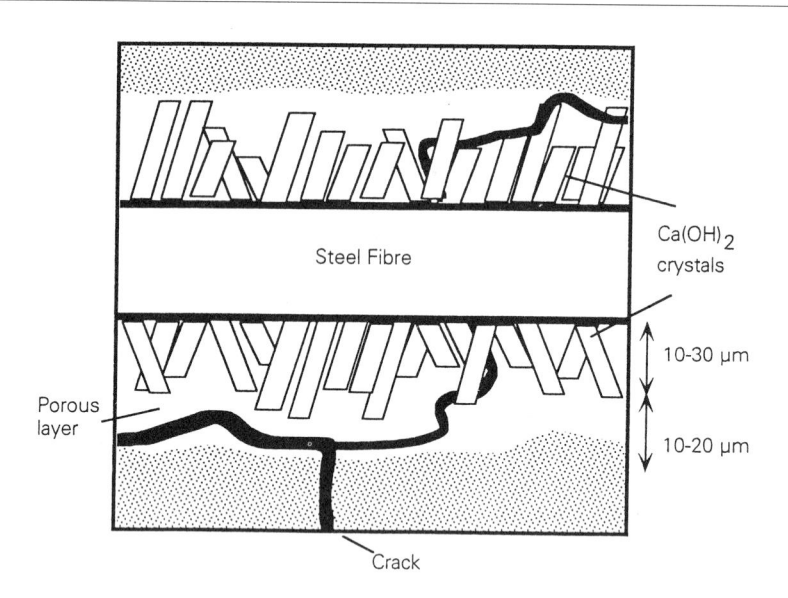

Figure 11.3 Schematic diagram of crack arrest and debonding in the porous layer in the transition zone

to the interface. This layer is formed during the setting of the cement when calcium hydroxide crystals are nucleated and grow from the steel surface to form a layer around the fibre. This layer is then surrounded by a porous cement layer. Observations have shown that a crack propagating through the cement matrix is apparently arrested at a distance of about 10 to 40 µm away from the fibre and then it is deflected along this porous cement layer. This is shown schematically in Figure 11.3.

As well as increasing the toughness and tensile strength of the cement, the incorporation of steel fibres has been shown to have beneficial effects in reducing the creep and shrinkage of the composite material (Beaudoin, 1990, p. 104).

Better results can be obtained by adding fibres to sprayed 'Gunite concrete' or 'Shotcrete' in place of reinforced steel nets or bars. The use of fibres as reinforcement instead of steel mesh or bars has several advantages, namely, savings can be made if there is no necessity for the placement of the reinforcement prior to the application of the shotcrete, there is a much greater flexibility of shape that can be sprayed, and there are no difficulties associated with ensuring an adequate uniform depth of coverage as when steel reinforcement has to be protected from corrosion of the rebars (Skarendahl, 1994).

Fibres are added to the wet- or dry-mix shotcrete in order to improve toughness and also to help reduce cracking while the cement is setting.

In the dry mix process, cement and sand are mixed with the fibres and then water is added just before the mix is sprayed. The wet mix utilizes premixed wet concrete to which fibres are added just prior to spraying.

The addition of steel fibre reinforcement in combination with microsilica to shotcrete results in a further improvement of properties such as a reduction of rebound of the mix and an increase in the strength of the fibre/cement bond. Morgan (1991) describes the use of steel fibres in shotcrete for the support of underground openings (tunnels) and compares the properties of the linings with those of shotcrete sprayed onto steel mesh reinforcement and plain shotcrete. The use of fibres in the shotcrete results in a large increase in its toughness. As well, it eliminates the need for the placement of reinforcing steel mesh prior to spraying.

11.1.2 Asbestos

Up to about 20 years ago, practically all commercial fibre-reinforced cement materials were produced with asbestos fibres as the reinforcing material. But it then became essential to find a substitute for the fibres, firstly because of the heavy consumption of asbestos ($\sim 20 \times 10^6$ tonnes/year), supplies of this mineral were becoming scarce, and secondly, because of the carcinogenic nature of the fibres (Krenchel, 1974).

The more common minerals that can occur as fibrous asbestos are:

Chrysotile (white asbestos) $3MgO.2SiO_2.2H_2O$

Amphiboles –   Crocidolite (blue asbestos)
$Na_2O.Fe_2O_3.3FeO.8SiO_2.H_2O$

        Amosite (brown asbestos)
$5.5FeO.1.5MgO.8SiO_2.H_2O$

        Anthophyllite $7Mg0.8SiO_2.H_2O$

        Tremolite $2CaO.5MgO.8SiO_2.H_2O$

        Actinolite $2CaO.4MgO.FeO.8SiO_2.H_2O$

(*Note*: The term asbestos refers to the crystal morphology and not the chemical composition. Asbestiform minerals consist of fibres with a l/d of >50:1. All of the amphibole minerals listed above occur in both fibrous and non-fibrous varieties of crystals.)

The morphology of chrysotile and the fibrous (asbestiform) amphiboles is shown in Figure 11.4.

The fibrous nature of these minerals is due to the very strong bonding within a crystalline unit, and weak bonding in between the units. The major difference between the chrysotile fibres and the asbestiform amphibole fibres is that the chrysotile fibres are hollow and this makes them

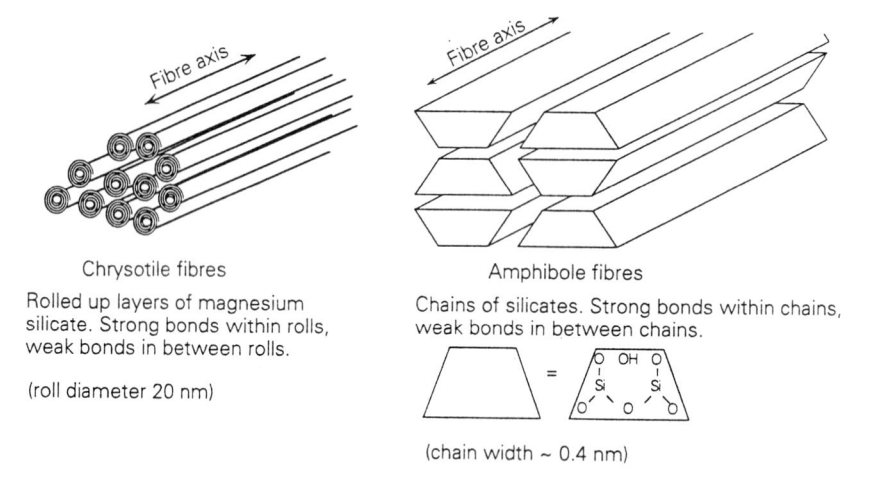

Figure 11.4 The morphology of the asbestos minerals

flexible and 'silky' in texture, whereas the amphiboles tend to be more rigid.

Asbestos fibres are chemically resistant to the alkaline cement environment, and also have good mechanical properties: high tensile strength (~1 GN/m^2), high modulus of elasticity (~150 GN/m^2) and a relatively low density (~3000 kg/m^3) (Krenchel, 1974). Because of the nature of the crystal structures of asbestos, the fibre length remains almost unaltered during the mixing of the cement, whereas the fibre bundles are split to form finer fibrous bundles.

The bond that is formed between the fibres and the cement paste is strong and there is no evidence of the formation of an interfacial zone of calcium hydroxide next to the fibres, as has been observed in other fibres used to reinforced cements. The actual toughening mechanism of these fibres is complex being dependent, in part, on the ability of the fibres to split into finer fibres while retaining their high strength and high modulus of elasticity. The high strength of the fibre/cement bond and the resistance of the fibres to chemical attack by the alkaline environment also contributes to the toughening of the cement. When these properties are combined with the ease of mixing of the composite, it is not surprising that finding safe substitutes for asbestos has proved to be a very difficult task, and compromises have had to be made.

Asbestos cement sheet was formed by a 'paper-making' process in which 8–15% asbestos was mixed with Portland cement and water. This mix was sprayed onto gauze and part of the water was then squeezed leaving a wet asbestos cement blanket. This was then transferred to rollers and built up to the required thickness. The sheet was then cut, removed from the rolls and more water was then removed. The cement was then rapidly cured by steam (Ryder, 1976).

Silica sand could also be added to the cement as a diluent. The resulting sheet was reasonably flexible, strong and light weight. The sheet could be made more flexible by the incorporation of organic (wood cellulose) fibres.

However, due to the problems in handling asbestos, asbestos cement sheeting is now no longer produced in many parts of the world and other fibres have replaced asbestos fibres in almost all applications.

11.1.3 Synthetic mineral fibres (glass fibres)

There is a large range of glass and synthetic mineral fibres available. They can be produced with very high tensile strength properties and good elastic properties, and have relatively low density (Krenchel, 1974).

E glass fibres (a borosilicate) are attacked by the high alkalinity of the cement. In order to minimize this they are normally coated with resin, which also serves to protect the fibres to some extent from mechanical damage during the mixing process.

E glass fibres are available either as long single filaments, which have to be chopped and bundled before mixing into the cement, or as mats of spun and blown fibres.

Although the spun and blown fibres are much cheaper than the fibres produced by the single filament process, the spun fibres tend to tangle into ever tighter knots during the mixing process (unlike the asbestos fibres which separate out during mixing). It is difficult for the cement paste to penetrate these tangles, and the result is an uneven distribution of the fibres. This leaves a large part of the reinforcement totally unanchored and extensive areas of unreinforced matrix material. The resulting product has poorer mechanical properties than the corresponding unreinforced cement material. Therefore it is necessary to use single-filament (~10 μm diameter) fibre material which has been chopped into suitable lengths. These fibres are manufactured by drawing the molten glass through a platinum bushing. The organic coating is applied to hold the individual fibres together into bundles or strands. Bundles of ~200 filaments are used. The bundle diameter is ~120 μm and length ~30 mm.

Special alkali resistant glass (AR glass) has been developed, by Owens-Corning, which contains ~16% ZrO_2. Pilkington have also developed an alkali resistant fibre and it is claimed that cements reinforced with this fibre are stronger and far more resistant to breakage than asbestos reinforced cements. However, the long-term behaviour of this composite is not yet known and Pilkington have placed an embargo on its full structural use except where a failure would not be disastrous.

The physical properties and chemical compositions of the two types of glass are shown in Tables 11.1 and 11.2.

Table 11.1 Properties of single glass filaments (Bentur and Mindess, 1990, p. 221)

Property	E glass	AR glass
Density (kg/m^3)	2540	2780
Tensile strength (MPa)	3500	2500
Modulus of elasticity (GPa)	72.5	70.0
Elongation at break (%)	4.8	3.6

Table 11.2 Chemical composition of glass filaments by mass %

	E glass (%)	AR glass (%)
SiO_2	52.4	71
$K_2O + Na_2O$	0.8	11
B_2O_3	10.4	–
Al_2O_3	14.4	18
MgO	5.2	–
CaO	16.6	–
ZrO_2	–	16
Li_2O	–	1

Glass-reinforced cement is used primarily for making panels, pipes, lamp posts etc. The 'spray-up' process is now widely used to produce thin, lightweight cladding panels for buildings. The chopped glass fibre and a cement slurry are simultaneously sprayed onto the form. (This method is very similar to that which can be used to produce fibre-reinforced polymers.) The cement slurry mixture has to be carefully controlled, water/cement ratios are <0.35 since above this ratio the physical properties of the panels are adversely affected. Water reducers and superplasticizers are often added. Compaction and the removal of entrapped air is achieved by the use of serrated hand rollers. For the production of thin cement panels, continuous glass fibres, woven into a mat can also be used. The mat must contain sufficiently large openings to allow the penetration of the cement grains (Bentur and Mindess, 1990, p. 216). Thicker sections of glass-fibre-reinforced cement are made by pouring rather than spraying. The fibres are added to the cement in the mixer, but care must be taken to avoid excessive damage to the fibres.

Regardless of the methods used to make the glass reinforced cements, the poor long-term behaviour of these composites remains the major problem restricting their use. The chemical attack of the cement on E

Table 11.3 The effects of aging on the glass fibre reinforced cement

Fibre type	Time (years)	Chemical degradation	Growth of CH
E glass	<1	Severe	Small
AR glass	<1	Nil	Nil
	5–40	Small	Large
	30–50	Large	Large

glass fibres, and the effect of AR glass fibres on the microstructure of the cement adjacent to the fibres are attracting a great deal of research. The use of AR fibres results in almost complete loss of toughness due to the nucleation and growth of massive calcium hydroxide crystals around the fibres. Table 11.3 summarizes ageing problems with glass-fibre-reinforced cement.

The life expectancy of E glass composites can be increased by completely sealing the fibres from the matrix, and/or by reducing of the alkalinity of the matrix. The life expectancy of AR glass composites can be increased by the modification of the nucleation and growth process of the hydration products around the fibres (Bentur and Mindess, 1990, p. 259).

Recently, the performance of glass-fibre-reinforced cements has been improved by the incorporation of microsilica or fly ash which reduce the alkalinity and eliminates the deposition of CH around the filaments. This was found to be particularly successful when the silica fume was present at the fibre/cement interface.

11.1.4 Carbon fibres

In recent times, carbon fibres have been used to reinforce cement. Their superior strength and stiffness characteristic, compared with steel, make them very attractive but the cost of the fibres is very high. It is essential that they be either coated or embedded in resin since the carbon fibres are even more vulnerable to surface damage than glass fibres (Zonsveld, 1976).

The fibres are 7–15 μm in diameter, and the high strength of the fibres is due to the preferred orientation of the layered planes of graphite along the fibre axis.

These fibres are manufactures by two processes, either from polyacrylonitrile (PAN carbon fibres) or from coal tar pitch or petroleum (pitch carbon fibres). Various types of fibres may be produced by these processes, the properties being dependent on the process details, such as heat treatment, oxidation and stretching. The PAN fibres are generally of higher quality, having a higher modulus and strength than the pitch fibres but the pitch fibres are very much cheaper than the PAN fibres. The lower price of the pitch fibres mean that these are normally the ones used in cement products. The fibre-reinforced cement sheet production methods are very similar to those used for glass-fibre reinforcement.

The pitch carbon fibres have a Young's modulus of 28–830 GPa and a tensile strength of 0.59–2.76 GPa (Beaudoin, 1990, p. 250). The fibres are from 3–10 mm in length and about 14 μm in diameter. Ohama *et al.* (1985) developed a method by which these carbon fibres could be dispersed in cement using an ordinary mortar mixer when silica fume and a water-reducing agent were added. They found that the strength, deformability, impact resistance, and drying shrinkage were all improved by increasing the carbon fibre content and they attributed this to the improved anchorage of the fibres caused by the presence of the silica fume additions.

11.1.5 Polymer fibres

(a) Nylon

Nylon, in the form of cut monofilaments, was the first polymer fibre to be recommended for the construction of blast-resistant buildings. It has not been extensively used because it was soon overtaken by the much cheaper polypropylene fibre, which has similar properties (Zonsveld, 1976).

(b) Polypropylene

The cheapest of the synthetic fibres used in cement is fibrillated polypropylene twine which has a unique 'net' structure, shown in Figure 11.5. This structure is obtained by the extrusion of polypropylene tape followed be

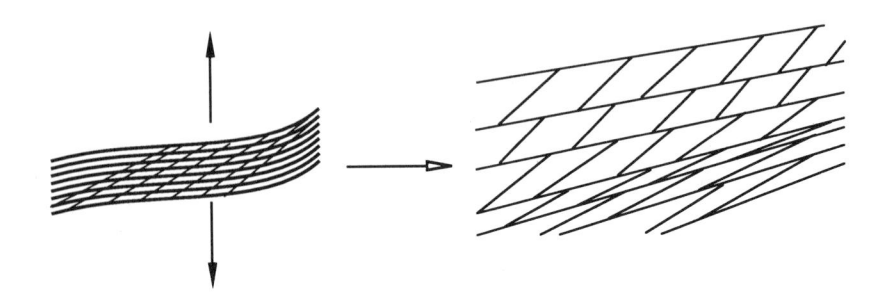

Figure 11.5 The structure of fibrillated polypropylene fibre produced by the splitting of a polypropylene tape

stretching and then twisting to form a yarn. When mixed with cement the fibres open up, allowing the cement mixture to penetrate the mesh and form a continuous phase in which the fibre is firmly trapped (Zonsveld, 1976).

An early problem with the use of polypropylene fibre was that the surface of the fibre was hydrophobic. This resulted in a very poor cement/fibre bond making it difficult to achieve a uniform dispersion of the fibres in the cement mix. It was therefore necessary to modify the surface of the fibres to improve their wetting properties, and various patented treatments are carried out to achieve this.

Polypropylene fibres are used in two ways to reinforce cement:

- As an alternative to asbestos to make thin sheets of fibre-reinforced cement. In this case the fibre content is normally greater than 5% (by volume). The properties of these sheets are dependent on the method of production (more details can be found in Bentur and Mindess, 1990, p. 325). In general the polypropylene reinforced material has a lower strength in tension than asbestos sheet, but the polymer composite has a higher toughness, and can withstand higher strains before fracture.
- To control cracking of cement or concrete. In this case much less fibre is used (up to 3% by volume), and its main function is as a secondary reinforcement. The fibres firstly control the cracking of fresh cement due to shrinkage, and secondly improve the toughness and strain capacity of set cement.

Polypropylene fibres in conjunction with fly ash in shotcrete have been used in Halifax (Canada) to seal and protect pyritic rock outcrops from leaching by rainwater and snow melt which, if it took place, would present serious health and environmental problems due to the aggressive nature of the leachates. The combination of fibres, fly ash, resin air entraining admixture with cement and aggregate resulted in a shotcrete that had satisfactory workability, drying shrinkage and toughness, combined with excellent freeze-thaw durability (Malhotra *et al.*, 1994).

(c) Aramid (Kevlar)

The main difference between Kevlar and carbon or glass fibres is that it fractures in a ductile manner (Krenchel, 1974, p. 363). The durability of the fibres has been found to be good, but there is some indication that alkali attack might be a problem. The strength retention of Kevlar and other fibres after immersion in alkaline solution are shown in Table 11.4.

One of the main advantages that the Kevlar fibre/cement composite has over the other polymer fibre/cement composites, is its resistance to heat. This means that, in the production of asbestos cement replacements, the Kevlar fibre-reinforced cement can be cured by autoclaving at up to 150 °C.

Table 11.4 Strength retention after 104 h in pH 12.5 solution (Bentur and Mindess, 1990, p.363)

Fibre	% original strength
AR glass	42.5
Carbon	40.5
Kevlar	90.0
Steel	99.6

(d) Polyvinyl alcohol (PVA)

PVA fibres are produced by either wet or dry spinning and again have been developed as an asbestos substitute. The hydrophilic nature of the surface of these fibres means that they can be dispersed efficiently in the cement and that they also form a strong fibre/cement bond with the hardened cement. It is stated that the mechanical properties of the PVA/cement composites are superior to those of asbestos cement (Bentur and Mindess, 1990, p. 369).

11.1.6 Natural fibres

With the demise of asbestos as a low-cost reinforcing material, there is an increasing use of natural fibres to produce low cost cement composite materials (Bentur and Mindess, 1990, Ch. 11). These fibres include jute, sisal, coconut and wood (cellulose). The microstructure of these fibres is very complex, and the properties of the composites vary widely. The durability of the composites are similarly affected by many processes, these include degradation due to alkaline and/or biological attack and petrification of the fibres which leads to embrittlement.

11.1.7 Summary of the properties of the fibres

Typical properties of the fibres used to reinforce cement (Krenchel, 1974) are shown in Table 11.5.

11.2 FIBRE-REINFORCED CONCRETE

The fibre material most commonly used for the reinforcement of concrete is steel. Polymer fibres have also been used in the production of blast-resistant structures, and some trials have been carried out on the use of glass fibres.

FIBRE-REINFORCED CONCRETE

Table 11.5 Typical properties of fibres

Fibre	Diameter (μm)	Specific gravity	Modulus of elasticity (GPa)	Tensile strength (GPa)	Elongation at break (%)
Steel	5–500	7.84	200	0.5–2.0	0.5–3.5
Glass	9–15	2.60	70–80	2–4	2–3.5
Asbestos					
Crocidolite	0.02–0.4	3.4	196	3.5	2.0–3.0
Chrysotile	0.02–0.4	2.6	164	3.1	2.0–3.0
Polypropylene	20–200	0.9	5–77	0.5–0.75	8.0
Kevlar	10	1.45	65–133	3.6	2.1–4.0
Carbon	9	1.90	230	2.6	1.0
Nylon	–	1.1	4.0	0.9	13.0–15.0
Cellulose	–	1.2	10	0.3–0.5	–
Acrylic	18	1.18	14–19.5	0.4–1.0	3
Polyethylene	–	0.95	0.3	0.0007	10
Wood fibre	–	1.5	71.0	0.9	–
Sisal	10–50	1.50	–	0.8	3.0
Cement matrix (for comparison)	–	2.5	10–45	0.0037	0.02

Theoretical studies on fibre-reinforced concrete (FRC) began in the 1950s, and dealt mainly with steel-fibre-reinforced concrete (SFRC). It is reported by Krenchel (1974) that, with the exception of asbestos, steel is the most common fibre used in the reinforcement and toughening of concrete.

The fibres are normally made of carbon steel, but where corrosion resistance is important, alloy steels are used. The methods used to produce the fibres are shown in Table 11.6.

Table 11.6 Methods of production of steel fibres for SFRC

	Fibre morphology	Method of manufacture
Round fibres		Cutting or chopping wires
Flat fibres		Shearing sheets or flattening wires
Crimped fibres		Deforming wires
Irregular fibres		Melt extraction (a rotating wheel is brought into contact with the surface of molten steel, and the frozen metal is thrown off to form fibres)

The strength of the fibres varies between 345 and 2100 MPa and is dependent on the production method used.

There are problems in mixing the concrete. It is difficult enough to mix steel fibres effectively in cement pastes, but difficulties are further increased by the added presence of aggregate in the concrete. The basic problem is to be able to introduce the required volume of fibres into the concrete and to disperse them uniformly. The fibres lower the workability of the concrete in proportion to their length and volume fraction. This is partially overcome by the use of deformed fibres, which can be shorter while producing the required increase in toughness. But even with these deformed fibres of length 12–50 mm, their volume fraction in the concrete is limited to about 1–2%. In the mixing there is a tendency for the fibres to clump together.

Both polymer and steel fibres can be mixed by adding loose fibres directly into the concrete mixer during the last part of the mixing period. Both types of fibres can tolerate the rough treatment this entails, but there is a tendency to tangle if the mixing is carried out for too long a time. This can be improved by premixing the fibres, cement and water, and then adding the aggregates.

It might be expected that the steel fibres would be susceptible to corrosion, particularly those near the surface of the concrete, and that this corrosion would lead to a decrease in the toughness of the concrete. However, this has proved not to be the case (Bentur and Mindess, 1990, p. 205). Even when the surface fibres have corroded, the effect on the toughness on the concrete has been found to be minimal. The general conclusion is that even though corrosion of fibres close to the surface results in the staining of the surface of the concrete, the corrosion products of the fibres are not sufficient in quantity to build up sufficient stress to crack the concrete matrix (Skarendahl, 1994). In marine environments, stainless steel or various grades of corrosion resistant steels can be used.

In cases where corrosion of the fibres has been observed, the mode of failure has been observed to change from fibre pull-out to fibre fracture.

The areas of application of SFRC seem to be limited only by the cost of the product. Since the addition of 1% steel fibre doubles the cost of the concrete, this limits the use of SFRC to special applications.

11.3 SOME USES OF FIBRE-REINFORCED CEMENT AND CONCRETE

11.3.1 Aircraft runways and other structural members

The thickness of a concrete runway slab can be reduced by a factor of two for the same wheel load and impact loading properties. Steel fibres have also been used for the reinforcement of frames, beams and flat slabs. It has been suggested that fibre reinforcement of concrete nuclear reactor pressure vessels could reduce the problems of the congestion of reinforcing rods, allow higher tensile stresses and provide better crack control.

11.3.2 Pipes

Glass-fibre-reinforced thin wall concrete pipes are being manufactured commercially in the UK. Continuous alkali-resistant glass-fibre reinforcement is used.

11.3.3 Blast and shock resistant structures

Here fibres (polymer) are used in conjunction with conventional reinforcement. They can also be used for the production of machinery beds that dampen vibrations and withstand shock loading.

11.3.4 Refractory applications

High alumina cement concrete reinforced with steel fibres can be used for high-temperature applications. For a hot face temperature of 1595 °C stainless-steel fibres are used, and for lower temperature (540 °C) plain carbon steel fibres are used. Such concretes have been used in linings for cement kilns and for the doors of open hearth furnaces. The resistance to the formation of large cracks and spalling is significant, even when the temperature of the hot face is sufficient to oxidize the fibres completely. The cracks propagating to the cooler side of the lining are resisted by the remaining fibres which are intact and have sufficient strength to prevent crack propagation through the lining.

11.3.5 Thin shells and walls

The use of fibres have resulted in the production of thinner and lighter walls for low cost housing and school buildings.

The Japanese have started to make use of carbon-fibre-reinforced concrete (CFRC) in the construction of high-rise office structures. Pitch-based fibres are used in concrete at 2 to 4% by volume (uniformly distributed throughout the concrete). The resulting concrete is about half the specific gravity of ordinary concrete, and produces structural members with 2 to 5 times the strength, one-half to one-third the cross section and one-fourth to one-tenth the weight of ordinary concrete (Akihama et al., 1988). This concrete was used in the construction of the 37-storey ARK Mori towers in Tokyo. The use of carbon fibres to reinforce the concrete curtain wall panels (total area 32 000 m^2, containing 170 tonnes of carbon fibres) resulted in a 60% reduction in wall mass, and the construction time was shortened. The decrease in mass also meant that a 2-tonne electric hoist could be used to position the panels instead of a tower crane. This concrete is as flexible as timber, can be nailed, planed and sawn.

Prior to the construction of the ARK building, the same material was used in the construction of the Al Shaheed monument in Iraq (Akihama *et al.*, 1988). This consists of two domes, each 40 m high with a base diameter of 45 m, consisting of a galvanized-steel skeleton covered with Turkish blue tiles. It was necessary that the cladding tile panels have a mass per unit area of less than 60 kg/m^2, because of the strength limit of the steel skeleton, and also be able to stand up to the climatic conditions in Bagdad (high summer temperatures and below freezing in winter). A normal Portland cement concrete curtain wall would have had a mass per unit area of 400 kg/m^2.

More recently, improvements have been made in the pitch based fibres, which have resulted in a doubling of the tensile and bending strengths of the composite material (Ford, 1990).

11.3.6 Marine applications

Fibre reinforcement has been used in the construction of breakwaters in order to improve the resistance of the concrete to wave impact. However there might be longer term corrosion problems. There is also the possibility that fibres might be used in conjunction with ferrocement for ship hull construction.

QUESTIONS

1. Why are short discontinuous fibres normally used in cement and concrete?
2. What is the essential role of these fibres?
3. Why should the term 'fibre-reinforced' not be used for cements and concretes that contain polypropylene fibres?
4. What is the difference in mechanical properties between glass-fibre-reinforced polymers and glass-fibre-reinforced cement?
5. Compare the behaviour of concrete reinforced with steel rebars, and concrete containing steel fibres.
6. Describe the role of the steel fibre/cement interface in the toughening of fibre-reinforced cement.
7. Why is asbestos cement sheeting is no longer produced in many parts of the world?
8. Explain the differences between the behaviour of steel, glass and asbestos fibres when they are mixed into wet cement.

9. Why is it necessary to coat glass fibres with polymer and then form them into bundles before mixing with cement?
10. Soda glass (window glass) is very much cheaper that E glass or AR glass. Why cannot it be used to reinforce cement?
11. What effect does silica fume additions have on carbon reinforced cement?
12. What limits the amount of steel fibres that can be dispersed in a concrete mix?
13. Why is corrosion of steel fibres in concrete not a major problem?
14. Describe some applications of steel, carbon and polymer fibre-reinforced cement or concrete.

References

Akihama, S., Suenaga, T. and Nakagawa, H. (1988) Carbon fibre reinforced concrete. *Concrete International*, **10** (1), 40–7.

Beaudoin, J. J. (1990) *Handbook of Fibre-Reinforced Concrete*, Noyes Publications, Park Ridge, N. J.

Bentur, A. and Mindess, S. (1990) *Fibre Reinforced Cementitious Composites*, Elsevier Applied Science, London; New York.

Ford, R. G. (1990) New developments in high performance cementitious materials. *Journal of the Australian Ceramic Society*, **26** (2), 163–70.

Krenchel, H. (1974) Fibre reinforced brittle matrix materials. In *Fiber Reinforced Concrete*, SP 44-3, American Concrete Institute, Detroit, pp. 45–77.

Malhotra, V. M., Cafette, G. G. and Bilodeau, A. (1994) Mechanical properties and durability of polypropylene fibre reinforced high volume fly ash concrete for shotcrete applications. *ACI Materials Journal*, **91**, 478–86.

Morgan, D. R. (1991) Steel fibre reinforced shotcrete for support of underground openings in Canada. *Concrete International*, **13** (11), 56–64.

Naaman, A. E. (1985) Fibre reinforcement for concrete. *Concrete International*, **7** (3), 21–5.

Ohama, Y., Amano, M. and Endo, M. (1985) Properties of carbon fibre reinforced cement with silica fume. *Concrete International*, **7** (3), 58–62.

Ryder, J. F. (1976) Applications of fibre cement in fibre reinforced cement and concrete. In A. Neville (ed.), *Fibre Reinforced Cement and Concrete*, Construction Press, Lancaster, pp. 23–35.

Skarendahl, Å. (1994) Developments and use of steel fibre shotcrete. In A. Aguado, R. Gettu, and S. P. Shah (eds), *Concrete Technology: New Trends, Industrial Applications*, E. & F. N. Spon, London, pp. 257–72.

Swamy, R. N., Mengat, S. and Rao, C. U. S. K. (1974) The mechanics of fibre reinforcement of cement materials. In *Fiber Reinforced Concrete*, SP 44–1, American Concrete Institute, Detroit, pp. 1–28.

Zonsveld, J. J. (1976) Properties and testing of concrete containing fibres other than steel. In A. Neville (ed.), *Fibre Reinforced Cement and Concrete*, Construction Press, Lancaster, pp. 217–26.

12 Deterioration and corrosion of concrete

Damage to concrete can occur by deterioration due to chemical or physical reactions taking place within the concrete itself or by the corrosion of reinforcement used in the concrete.

12.1 DETERIORATION OF CONCRETE

This generally follows reactions in the concrete due to chemicals present in the original mix or exposure of the concrete to environmental influences such as soft water or sea water.

12.1.1 Soft water

The deterioration of concrete exposed to soft water, for example pure mountain water, is described in detail by Biczók (1967, pp. 149–61). Soft water contains only small amounts of dissolved calcium or magnesium ions, and generally is slightly acidic having a pH < 7. The aggressiveness of soft water to concrete is dependent on the degree of hardness and the amount of 'free' carbon dioxide present in the water. The degree of hardness is a measure of the amount of calcium carbonate in the water, but the actual numerical values given are dependent on the standard used. For example, in the British system, one degree of hardness is one part of calcium carbonate (by mass) per 80 000 parts water, whereas in the French system, it is one part of calcium carbonate per 100 000 parts of water. 'Free' carbon dioxide is carbon dioxide dissolved in water to form carbonic acid.

Deterioration of concrete submerged in soft acid water involves firstly the water leaching species out of the concrete that are responsible for the maintenance of the high pH in the concrete. This, in turn, can lead to the decomposition of the other constituents in the concrete which are only stable at high pH, namely the hydrosilicates, hydroaluminates and hydroferrites which leave behind calcium carbonate filled frameworks of silica, alumina and iron oxides (Lees, 1992). The high pH in the concrete is mainly due to the **CH** and the **CSH** gels. However not all of the **CH** present in the concrete is free to maintain the pH at high levels as some could be encapsulated by other insoluble hydration products. (Biczók (1967, p. 151) considers that concretes based on Portland cement deteriorate if the calcium hydroxide content is reduced by more than 20%.) The pH of concrete is also affected by the solution of **CSH** gels in water since the CaO is selectively leached out of the gel which has the effect of maintaining the pH at high levels until the Ca/Si ratio in the gel is significantly reduced (Taylor, 1990, p.153).

Leaching occurs not only from submerged concrete, it can also be a problem in concrete exposed to rain water.

The extent of leaching in soft water is influenced by many factors which include the following.

(a) The hardness of the water

Hard waters contain high concentrations of dissolved calcium and magnesium ions which reduce the extent of the leaching of the calcium hydroxidefrom the concrete. The presence of these ions in the water reduces its aggressiveness because they react with the free carbon dioxide dissolved in the water to form carbonates.

(b) The flow rate of the water over the concrete

In stagnant water, or very low water flow rates, the layer of water in contact with the concrete becomes saturated with calcium hydroxide and this reduces to a low value the rate of its diffusion into the water from the concrete. If the water flow rate is high, then leaching of the calcium hydroxide from the concrete will not diminish.

(c) The amount of water seepage through the concrete

Soft water seeping through retaining walls, dams or water tanks dissolves the calcium hydroxide as it passes through the concrete. and emerges with increased calcium hydroxide content. As the water emerges the dissolved calcium hydroxide reacts with carbon dioxide in the air and precipitates out on the surface of the concrete in the form of white crystals of calcium carbonate (efflorescence). These precipitates are often referred to as 'white death' as they are indicative of concrete deterioration due to leaching.

(d) Additions of pozzolanas or blast furnace slag

Concretes that are made with blast furnace slag or pozzolana additions are more resistant to attack by soft water, firstly because some of the calcium hydroxide produced by the hydration reactions of the cement has been removed by the hydration reactions of the additives, and secondly because these hydration reactions (which produce **CSH**) lead to a reduction of the permeability of the concrete.

(e) The density of the concrete

The higher the porosity of the concrete, the more access the soft water has to leach the calcium hydroxide from the concrete and therefore the greater the diffusion rate of leaching. Leaching itself causes an increase in the porosity of the surface layers, and this progressively increases the access of water to the bulk of the concrete.

(f) Quality and condition of the concrete surface

During the ageing of concrete, the calcium hydroxide within the surface layers is gradually converted to calcium carbonate by reaction with carbon dioxide from the air. The formation of insoluble calcium carbonate within the surface layer serves to protect the underlying soluble calcium hydroxide from leaching by soft water. Sealing layers, such as bituminous paints, are also often applied. The concrete surface can be improved by the application of a cement mortar to produce a less porous smooth surface.

(g) Presence of sodium chloride

If sodium chloride is also present in soft water, then the attack is accelerated, as sodium chloride increases the solubility of calcium hydroxide and other compounds in the concrete.

12.1.2 Sulphate attack

Concretes which have been only mildly attacked by sulphates are whitish in appearance, more severe attack causes the set concrete to expand which leads to cracking and spalling. Ultimately the concrete becomes friable and is reduced to a soft mud. The strength of the concrete might actually increase during the early stages of attack, but this is followed by a loss of strength that precedes the overall expansion (Lawrence, 1990).

As was briefly described in Chapter 3, the major cause of the sulphate attack is the reaction of gypsum (calcium sulphate, $C\bar{S}H_2$) with hydrated

DETERIORATION OF CONCRETE

compounds in the set cement to form ettringite ($C_6A\bar{S}_3H_{32}$) which results in expansion and cracking of the set concrete (Cohen and Bentur, 1988). Some possible reactions are shown below:

$$3C\bar{S}H_2 + C_3A + 26H \rightarrow C_6A\bar{S}_3H_{32}$$

$$2C\bar{S}H_2 + C_4A\bar{S}H_{12} + 16H \rightarrow C_6A\bar{S}_3H_{32}$$

$$3C\bar{S}H_2 + C_4AH_{13} + 14H \rightarrow C_6A\bar{S}_3H_{32} + CH$$

Sodium and magnesium sulphates (**N$\bar{S}$, M$\bar{S}$**) can cause sulphate attack because they can initially react with calcium hydroxide (**CH**) (which is present in the set cement formed by the hydration of **C$_3$S** and **C$_2$S** reactions) as shown below:

$$N\bar{S} + CH + 2H \rightarrow C\bar{S}H_2 + NH$$

$$M\bar{S} + CH + 2H \rightarrow C\bar{S}H_2 + MH$$

where **N** = **Na$_2$O** and M = MgO.

Potassium sulphate behaves in a similar manner to sodium sulphate.

The gypsum formed by these reactions then reacts with hydrated compounds to form ettringite as before.

The attack by magnesium sulphate is particularly damaging because, as well as forming sparingly soluble magnesium hydroxide, which forces the above reaction to the right to form gypsum, **M$\bar{S}$** will also react with the **CSH** gels present in the set cement to form more gypsum (Cohen and Bentur, 1988):

$$xM\bar{S} + C_xS_yH_z + (3x + 0.5y - z)H \rightarrow xC\bar{S}H_2 + xMH + 0.5yS_2H$$

The gypsum formed in this reaction will also react with the calcium aluminates. As well as this, the magnesium hydroxide produced in the reaction with the **CSH** gels, together with that produced by the reaction of magnesium sulphate with calcium hydroxide, can combine with the silica hydrate (**S$_2$H**), produced by the reaction with the cementitious gels, to form a noncementitious product (**M$_4$SH$_{8.5}$**) (Cohen and Bentur, 1988).

Thus the presence of magnesium sulphate not only produces gypsum from the reactions with calcium hydroxide which in itself is an expansive reaction (Lees, 1992) with a theoretical volume expansion of 2.2, but also reacts with the **CSH** gel to form ettringite which causes further expansion and cracking of the set cement. In addition to these expansive reactions magnesium sulphate destroys the cementing effect of the **CSH** gel by the replacement of the Ca^{2+} ions in the gel by Mg^{2+}.

Sodium sulphate is not as damaging as magnesium sulphate as it cannot react with the **CSH** gel; this is because the Na^+ ions cannot replace the Ca^{2+} ions in the gel.

Table 12.1 Typical composition of sea water (g/l)

Na+	11.0
K+	0.40
Mg^{2+}	1.33
Ca^{2+}	0.43
Cl–	19.80
SO_4^{2-}	2.76

12.1.3 Concrete in sea water

Deterioration of concrete exposed to sea water occurs by the chemical action of dissolved salts in the sea water. The composition of sea water in the oceans (shown in Table 12.1) is remarkably constant throughout the world except for isolated seas such as the Dead Sea. The pH of surface sea water is also reasonably constant being between 8.0 and 8.2.

The chemical action of sea water on concrete was first recognized by Vicat in 1812 and is mainly due to attack by $MgSO_4$. The mode of the attack is complicated by the presence of chlorides in the sea water which retard the swelling that is normally characteristic of attack by magnesium sulphate. Concrete deterioration is apparently due to the loss of part of its constituents: calcium hydroxideand calcium sulphateare more soluble in sea water than in fresh water and are more readily leached out of the concrete.

The hydrated **CSH** in set cement is decomposed by magnesium sulphate to form hydrated silica, gypsum and magnesium hydroxide, but, in dense concrete the deposition of the magnesium hydroxide can block pores in the cement and slow down the attack. In general, sulphate attack is not a major problem for concrete in sea water. A minimum level of **C_3A** is desirable (normally 5–6%) as it reacts with the chloride ions to form complex salts and this serves to reduce the mobility of these ions through the concrete. This means that the use of sulphate resisting cement, which typically has a **C_3A** content of less than 5 wt%, should be avoided because it results in an increase in the permeability of the concrete to chloride ions (Gerwick, 1988).

For concrete structures in sea water, the splash zone is the most vulnerable to damage. It not only suffers from the concentration of salts under conditions of alternate wetting and drying, but also from mechanical damage from wave impact and abrasive damage from the impact of sand, gravel, rocks and ice. The temperature of the sea water also affects its attack on concrete. Deterioration of concrete by chemical attack is more severe in warm climates whereas the mechanical effects of alternate freezing and thawing of the water in the concrete is a major problem in cold climates. In warm climates marine borers (rock-eating molluscs) have

DETERIORATION OF CONCRETE

caused problems. The acid produced by the borers attacks limestone aggregates which results in holes being bored in the concrete through to the steel reinforcement (Gerwick, 1988). Direct access of the sea water then quickly corrodes the reinforcement.

12.1.4 Sulphuric acid attack

Sulphuric acid is the most likely of the mineral acids (hydrochloric, nitric, and sulphuric) to be found in natural ground water. It can be generated by the oxidizing action of aerobic bacteria (thiobacillus ferrioxidans) on sulphide minerals such as FeS. The acid will react with the surface of set Portland cement to form gypsum from the calcium hydroxide in the cement. This attacks the concrete foundations of structures set in the ground.

Another area of attack by sulphuric acid occurs in concrete sewage pipes. In this case, if the sewage system is poorly designed and the sewage is stagnant or flowing very slowly in the pipes, the action of anaerobic bacteria (desulphovibrio desulphuricans) can produce hydrogen sulphide gas. The gas dissolves in moisture above the water line, where aerobic bacteria, such as thiobacillus thioxidans, convert it to sulphuric acid which then attacks the concrete. Thus it is essential to keep sewage flowing in the system at such a rate that there is insufficient time for the anaerobic bacteria to act (Harrison, 1987).

Sea water bacteria can attack concrete, near the waterline, by the generation of weak sulphurous acid. One of these anaerobic bacteria is named thiobacillus concretivorous! (Gerwick, 1988).

12.1.5 Alkali-silica reaction

The alkalis involved in the alkali-silica reaction are the hydroxides of sodium and potassium. These alkalies are often present in the raw materials used to make cement and after hydration are present mostly in soluble form (Gerwick, 1988). Most discussions refer to the alkali as sodium hydroxide, but in actual cases the term alkali should refer to either sodium or potassium hydroxide or both.

Fine-grained, porous siliceous aggregates are reactive and potentially dangerous. They have a large surface area on which the reaction can take place between the alkali and aggregate material. The amount of alkali that is available for the reaction normally is dependent on the impurity levels of alkali containing compounds in the raw materials used to make the cement. If this is high, and high-surface area siliceous aggregate materials are used, then attack (reaction) can occur. In this case, the product of reaction is a sodium or potassium-rich silicate gel. It is capable of absorbing large amounts of water which causes the gel to swell. If this gel completely fills the pores and voids in the concrete, then any further swelling results in a pressure build-up which cracks the concrete. If the gel then dehydrates, the cracks formed are left open.

The cracking that is associated with the alkali–silica reactions occurs in the bulk of the concrete, since the pressure build-up of reactions that occur near the surface is accommodated by the gel exuding to the surface where the pressure is relieved (Buck, 1987).

In general, non-porous siliceous aggregates, such as quartz, do not cause problems due to the attack of alkalis because the extent and rate of attack is low. However, strained microcrystalline quartz does show reactivity. According to Buck (1987) such strained quartz can be identified by the optical microscopic measurement of the undulatory extinction (UE) angle of the quartz. This identification is still uncertain, Gratten-Bellew (1987) considers that the correlation between the UE angles and rates of expansion due to the alkali–silica reaction observed with microcrystalline quartz might be fortuitous.

It appears that the properties of the gel that is formed are dependent on the alkali/silica ratio of the gel. There is a critical value for this ratio, below which the gel is non-swelling. Therefore if there is a large surface area of silica (for example, microsilica in the concrete), or if there is a low concentration of alkali, non-swelling gels will be formed and there will be no damage to the concrete.

The concept of the formation of swelling and non-swelling gels, depending on the relative concentrations of alkali and silica, is not necessarily the whole answer to the observed effects. Other investigations have indicated that there is competition for the alkali between an active aggregate and microsilica additions for the alkali (Wang and Gillott, 1992). The argument put forward in this case is that the small size and the dispersed nature of the silica gel result in only a small swelling pressure around the microsilica particles and that this can be accommodated by the surrounding cement paste without causing any measurable expansion, whereas the attack on the relatively massive opal aggregate particles would result in expansion because of the large amount of reactive silica in the aggregate. It is also possible that the alkali attack on the microsilica is completed so quickly that it occurs before the cement paste has hardened. This suggests that, in the presence of sufficient microsilica, the alkali–microsilica reaction will compete with the alkali–aggregate reaction and consume the available alkali before damage can be done.

In summary, several conditions have to exist for alkali/silica reactions to occur that will result in cracking of the set concrete. There must be a sufficiently high concentration of sodium or potassium hydroxide in the cement, there must be some reactive silica material present but not so much that the non-swelling gel will form, there must be a source of water for the gel to absorb and the porosity of the concrete must be too small to accommodate the swelling gel.

DETERIORATION OF CONCRETE

12.1.6 Alkali-carbonate reaction

Concretes made with a dolomitic limestone ($CaCO_3.MgCO_3$) aggregate can suffer from reaction of this aggregate with alkali which results in the loss of bond strength and microcracking of the concrete (Regourd, 1984). The sequence of the reaction is as follows:
reaction of dolomite with the alkali, XOH, where X = Na, K or Li to form alkali carbonate

$$CaCO_3.MgCO_3 + 2XOH \to Mg(OH)_2 + CaCO_3 + X_2CO_3$$

and then reaction of the alkali carbonate with calcium hydroxide to reform the alkali.

$$X_2CO_3 + Ca(OH)_2 \to 2XOH + CaCO_3$$

This results in the regeneration of the alkali for further attack on the aggregate.

12.1.7 Frost attack

Successive freeze/thaw cycles of water in concrete can result in damage to the concrete. The surface of the concrete exposed to air temperature swings is normally that affected, and the general result of the freeze/thaw cycles on concrete roadways is often delamination of the road surface (Mallett, 1994).

The exact mechanism by which the action of frost damages concrete is complex, but two causes can be identified. The first cause is by the freezing of water in the capillary pores in the concrete. The capillary pores are those that are present in ordinary strength concrete due to the necessity of having to add more water to the concrete than is necessary for the complete hydration of the cement in order to make the concrete workable. It is more difficult for this water to freeze than open water due to the presence of dissolved salts which lower the freezing point. Nevertheless freezing will occur eventually and, as the water freezes, it expands by about 9%. The increase in hydraulic pressure brought about by this expansion can result in crack formation.

The concentration of dissolved salts in the smaller gel pores (associated with the **CSH** gel) is higher than that in the capillary pores and hence the freezing point is lower than that of the water in the capillary pores. In addition, the liquid in the pores is under high surface tension forces and these prevent the formation of ice nuclei which are necessary for freezing to occur (Lees, 1992). As soon as some of the water freezes in the capillary pores to form ice (which is almost pure) the concentration of dissolved salts in the remaining water in the capillary pores increases and eventually becomes greater than that in the gel pores. This causes an increase in osmotic pressure between the capillary and gel pores and an osmotic flow of water from the gel to the capillary pores where it can freeze and further increases the hydraulic pressure.

Frost attack can be reduced either by the reduction of the volume of capillary pores in the concrete by using a lower water/cement ratio, or by the deliberate entrainment of air. The use of air entraining agents, as described in Chapter 7, results in the formation of a high concentration of minute air bubbles in the concrete. These bubble can serve to relieve the hydraulic pressure that results from the ice formation within the capillary pores, and hence prevents the expansion and cracking of the concrete. In fact, the entrainment of 10% air can give rise to an overall small contraction of the concrete on freezing. This is thought to be due to the effect of the dehydration of the gel pores by osmotic flow which results in a shrinkage of the **CSH** gels (Mehta and Monteiro, 1993, p. 131).

12.2 CORROSION OF STEEL REINFORCEMENT

A detailed description of the mechanism of corrosion of steel in concrete is beyond the scope of this book; only an outline will be given here. In general, the high pH of the concrete (~12.5), due primarily to the presence of calcium hydroxide, is sufficient to maintain the steel in a passivated state and corrosion will only occur when these conditions are changed and the pH falls.

For corrosion to occur three conditions must be met, namely: that part or all of the steel must be depassivated; that there must be an ionic path through the concrete, and that oxygen must be present. This is shown schematically in Figure 12.1

The corrosion of the steel reinforcement results produces iron hydroxides and oxides, which occupy a greater volume than the steel from which they are formed. It is this increase in volume which causes the concrete to crack and spall.

12.2.1 Depassivation of steel in concrete

The pH of concrete can be reduced to a level at which the steel is no longer in a passive state by carbonation of the calcium hydroxide. This can take place if carbon dioxide from the atmosphere can penetrate into the concrete and react with the calcium hydroxide to form calcium carbonate. Obviously the rate of this reaction will depend on the permeability of the concrete. Carbonation is not damaging to the concrete itself, but it causes a reduction in the pH of the concrete and, consequently, could lead to depassivation of the steel. This depassivation of the steel, when coupled with the penetration of oxygen and water, will promote oxidation

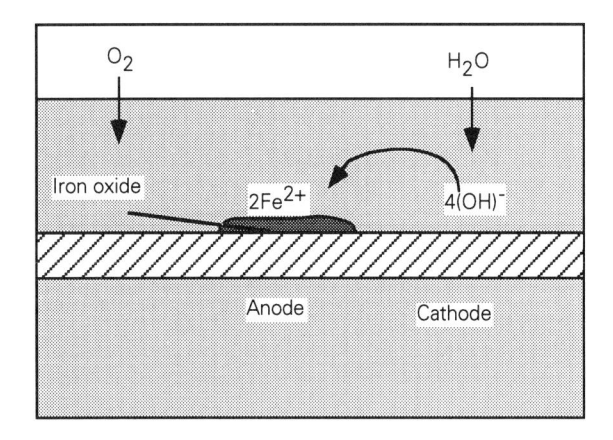

Figure 12.1 Conditions necessary for corrosion of steel reinforcement in concrete

(corrosion) of the steel, to form iron oxides. Under dry atmospheric conditions (relative humidity <50%) there will be no corrosion of the steel even if depassivation has occurred by carbonation. The lack of moisture means that there is no ionic path through the concrete. The penetration of water into the concrete is not only important in providing an ionic path through the concrete, but it also affects the rate of carbonation. If the concrete is dry then although carbon dioxide can diffuse through the capillary pore network, carbonation of the calcium hydroxide cannot occur as this reaction requires the presence of moisture in order to take place. On the other hand, if the concrete is saturated with water, then carbonation does not occur to a large extent because the diffusion rate of carbon dioxide in water is very low (about 10 000 times lower than in air). The most dangerous situation is where the pores are coated with a film of water. This allows the rapid penetration of carbon dioxide through the gaseous phase and also provides the moisture necessary for carbonation. This situation is likely to exist near the surface of the concrete at relative humidities greater than 70% (Bakker, 1988). The surface carbonation of cracks in concrete is particularly dangerous because if the cracks extend from the surface into the concrete to the reinforcement, the reinforcement is this area will be depassivated, become anodic and corrode.

Chloride ions in concrete can also result in the depassivation of steel reinforcement even when the pH is maintained at high values. This corrosion is often intense and localized (pitting corrosion) because corrosion cells are set up with small anodic areas activated by the chloride ions and large cathodic areas of passivated steel. The source of the chloride ions may be from the water used to mix the concrete, from the addition of calcium chloride accelerators or from the diffusion of chloride containing water into the concrete. The first two sources can be easily prevented, but the third is more difficult to control, particularly when the concrete is in contact with sea water, or when salt is used as a de-icing mixture on concrete roadways. In the latter cases, the best protection is to reduce the permeability of the concrete, or to try to reduce the mobility of the ions by chemical means. It is thought that C_3A will bind some chloride ions to form the compounds ($3CaO.Al_2O_3.CaCl_2.10H_2O$) (Friedel's salt) and $3CaO.Al_2O_3.3CaCl_2.32H_2O$ (Hartl, 1984).

As with corrosion due to depassivation by carbonation, the presence of both oxygen and moisture is essential for pitting corrosion to be sustained but if chloride ions are present they have a hygroscopic effect and the condensed moisture aids corrosion (Bakker, 1988).

Whether the depassivation and subsequent corrosion is dominated by carbonation or chloride ions depends on the situation. If there is no chloride present, then the steel will only corrode when carbonation reaches the reinforcement, whereas in the presence of chloride ions, corrosion will occur before full carbonation of the concrete covering the reinforcement has taken place.

12.2.2 Moisture and air

As described previously, for corrosion of embedded steel to occur both moisture and oxygen must be present. This means that corrosion of reinforcement is more likely to take place in concretes that have been made with high water/cement ratios, which result in the production of porous concrete due to the presence of excess water in the mix. Oxygen and water vapour can then diffuse through these pores and water can coat the pore surface.

12.2.3 Relationship between corrosion and tensile stress

Steel in tension will corrode at a higher rate than unstressed steel. Reinforcement steel will be in tension when loaded. Prestressed steel reinforcement is also always in tension. Since the integrity of the structure is dependent on the correct functioning of the prestressed members, it is essential that steel in prestressed concrete be given a high degree of protection in order to prevent its corrosion and possible failure.

12.2.4 Corrosion of steel fibres in fibre-reinforced cement and concrete

It has been observed that steel fibres rust only at the exposed surface of fibre-reinforced concrete, and that the amount of surface oxidation is minimal since the fibres make up only about 1% of the concrete (Vondran, 1991). This surface oxidation has no effect on the structure strength or integrity.

CORROSION OF STEEL REINFORCEMENT

The corrosion behaviour of the fibres is different from that of reinforcing bars, in that the bars extend throughout the structure and corrosion cells will be set up if the environment in which the bars are situated differs from one point to another. It is much more probable that this will be the case for reinforcement bars that extend over large distances than for short discontinuous fibres. The damage done by the corrosion of reinforcement bars is due to the production of lower density corrosion products which cause an expansive stress within the concrete which then can result in cracking and spalling of the concrete. If fibres do corrode, there is insufficient corrosion product produced from the small fibres to exert sufficient stress to cause failure and corrosion does not lead to catastrophic loss of strength of the concrete. The fibre-reinforced concrete can be used successfully in corrosive environments such as sea walls, bridge decks exposed to de-icing chemicals, and concrete breakwaters (Vondran, 1991).

QUESTIONS

1. Explain why 'soft' water is much more aggressive to concrete than 'hard' water.
2. Why is attack by magnesium sulphate potentially more damaging to concrete than attack by sodium sulphate?
3. Describe the alkali-silica reaction.
4. Why does the addition of fine active silica particles (microsilica) to concrete prevent the damaging alkali-silica reaction?
5. How can frost attack on concrete be reduced? Would you expect high-performance concrete to be more or less susceptible to frost attack than normal strength concrete?
6. Why is corrosion of steel in concrete damaging to the concrete?
7. Describe the conditions necessary for corrosion of steel reinforcement in concrete to occur.
8. What is meant by depassivation of steel in concrete, and how can it occur?

References

Bakker, R. F. M. (1988) Initiation period in corrosion of steel in concrete. In P. Schiessl (ed.), *Corrosion of Steel in Concrete*, Chapman & Hall, London; New York, pp. 22–55.

Biczók, I. (1967) *Concrete Corrosion and Concrete Protection*, Chemical Publishing Co., New York.

Buck, A. D. (1987) Petrographic criteria for recognition of alkali-reactive strained quartz. In P. E. Grattan-Bellew (ed.), *Concrete Alkali-Aggregate Reactions*, Noyes Publications, Park Ridge, N. J., pp. 419–23.

Cohen, M. D. and Bentur, A. (1988) Durability of Portland cement-silica fume pastes in magnesium sulphate and sodium sulphate solutions. *ACI Materials Journal,* **85**, 148–57.

Gerwick, B. C. Jr. (1988) Concrete marine structures. In D. F. Hanson and C. R. Crowe (eds), *Materials for Marine Systems and Structures*, Academic Press, Boston, pp. 352–87.

Gratten-Bellew, P. E. (1987) Is high undulatory extinction in quartz indicative of alkali-expansivity of granitic aggregates? In P. E. Grattan-Bellew (eds), *Concrete Alkali-Aggregate Reactions*, Noyes Publications, Park Ridge, N. J., pp. 434–9.

Harrison, W. H. (1987) Durability of concrete in acidic soils and waters. *Concrete,* **21** (2), 18–24.

Hartl, G, (1984) Physical processes related to concrete. In S. Rostam (ed.), *Durability of Concrete Structures*, Department of Structural Engineering, Technical University of Denmark, Lyngby, pp. 179–227.

Lawrence, C. D. (1990) Sulphate attack on concrete. *Magazine of Concrete Research,* **42** (153), 249–64.

Lees, T. P. (1992) Deterioration mechanisms. In G. Mays (ed.), *Durability of Concrete Structures*, E. & F. N. Spon, London: New York, pp. 10–33.

Mallett, G. P. (1994) *Repair of Concrete Bridges*, Thomas Telford, London, p. 157.

Mehta, P. K. and Monteitro, P. J. M. (1993) *Concrete: Structure, Properties and Materials*, 2nd edn, Prentice-Hall, New Jersey.

Regourd, M. (1984) Durability: physio-chemical and biological processes related to concrete. In S. Rostam (ed.), *Durability of Concrete Structures*, Department of Structural Engineering, Technical University of Denmark, Lyngby, pp. 49–71.

Schießl, P. (1984) Protection of reinforcement. In S. Rostam (ed.), *Durability of Concrete Structures*, Department of Structural Engineering, Technical University of Denmark, Lyngby, pp. 241–97.

Taylor, H. F. W. (1990) *Cement Chemistry*, Academic Press, London.

Vondran, G. L. (1991) Applications of steel fibre reinforced concrete, *Concrete International,* **13** (11), 44–9.

Wang, H. and Gillott, J. E. (1992) Competitive nature of alkali-silica fume and alkali-aggregate (silica) reaction. *Magazine of Concrete Research,* **44** (161), 235–9.

Durability and protection of concrete — 13

13.1 DURABILITY

Concrete is a very durable material, as exhibited by the number of Roman concrete constructions which survive to this day. To achieve this durability, the correct selection of materials, good design and strict quality control when mixing, placing and curing the concrete is essential.

In the previous section, the various factors that can adversely affect the durability of concrete were outlined. Deterioration of concrete structures can be caused by chemical agents such as soft water, sea water, sulphates, acids, alkalis and carbon dioxide. In general, attack by chemical agents and the corrosion of reinforcement is worsened by the ingress of the corrosive media through interconnected pores in the concrete. In addition to this, corrosion of the metal reinforcement within the concrete is a major cause of failure of concrete structures.

Physical causes of deterioration include freeze/thaw cycles, while differential expansion/contraction of the concrete components due to temperature changes was covered in an earlier chapter.

Careful selection of the correct type of cement, aggregate, the mix proportions and the water/cement ratio for a particular application can increase the durability of concrete structures. In order to make the selection it is necessary to investigate the site where the concrete is to be placed to identify, in advance, the existence of any potential hazards.

13.1.1 Mix design for durable, good quality concrete

The term 'good quality' is used to describe a concrete that can be placed easily, and which on setting produces a low porosity concrete.

Factors which should be taken into account in the design of the concrete mix include the type of cement to be used, the cement aggregate ratio, the type of aggregate, aggregate size distribution and the water/cement ratio. All of these will be dependent on the physical and chemical conditions to which the concrete is to be exposed. The design of the mix should be mainly aimed at the reduction of interconnected porosity which results from the use of too high a water/cement ratio. The elimination of microcracks in the concrete should also be considered. These cracks can originate from the entrapment of bleed water under aggregate and reinforcement, from drying shrinkage of the cement paste and from differences in the thermal expansions of the aggregate, paste and reinforcement. The presence of these cracks can reduce the resistance of the concrete to fatigue, abrasion and corrosion.

As well as the design of the mix, the placement of the concrete should be carefully supervised. The thickness of the concrete cover over reinforcement is of great importance in controlling the penetration of carbon dioxide or chloride into the concrete. This is because the depth of penetration is approximately a function of the square root of the exposure time. This means that halving thickness of concrete through which it would take 100 years for carbonation to reach the reinforcement would result in the carbonation reaching the reinforcement in only 25 years. Compaction and curing are also crucial. Poor compaction can result in greater than a ten-fold increase in permeability (Schießl, 1984). Bad curing, in which the concrete is allowed to dry out prematurely, can result in a five to ten-fold increase in the permeability of the concrete.

Pomeroy (1986) considers that the requirements for good quality concrete are an adequate cement content, good compaction followed by a thorough curing – the three Cs. When asked to assess the relative importance of these he suggested the ratio of 1:3:5 respectively.

The general factors that are needed to produce durable concrete were summarized by Pomeroy (1986) as:

- The use of good-quality concrete, with adequate cement content, good compaction and curing. The mixes on site must be of uniform consistency, be easily placed and not prone to segregation.
- The prevention of corrosion of steel reinforcement. It must be covered with an adequate thickness of good quality concrete. Steel reinforcement must not move during the pour, and must not act as a screen for aggregate material in the concrete.
- Consideration of maintenance. This must be considered at the design stage and scheduled in advance. For example the scheduled unblocking of drainage channels to prevent saturation of the concrete with water.

13.1.2 The use of curing compounds

To ensure adequate curing of concrete, curing compounds can be used to restrict the rate of water loss by evaporation from the surface of the concrete. These compounds form a continuous film over the surface of the concrete and are applied when the concrete has hardened sufficiently so that there is no damage to the surface. Application is normally carried out by spraying, brushing or rolling. Four classes of compounds are described in detail Appendix D of Australian Standard AS 3799–1990. Class A are wax based compounds either emulsified in water or dissolved in a solvent, Class B are resin based compounds, Class C are synthetic-based compounds and Class D are 'others' (e.g. PVA). As well as ensuring good curing of the concrete by the restriction of evaporation, reflective pigments, such as titanium dioxide or aluminium flake can be mixed with the curing compounds to reduce heating of the concrete by solar radiation.

13.1.3 The use of pozzolanas

Additions of pozzolanic materials to the concrete results in a decrease in porosity due to the formation of **CSH** gel with calcium hydroxide. However, these additives also result in an initial decrease in the rate of hardening of the concrete, even though the long-term strength is often higher than that of concrete without additions. This means that the curing of concrete which contain pozzolanic materials is even more critical than that of ordinary concrete (Schießl, 1984).

The improvement of concrete that incorporates pulverized fuel ash or other pozzolanas in concrete is due to the reduction in pore size in the concrete, which reduces its permeability. The resistance to sulphate attack is improved due to the reduction of the amount of **CH** available to react with the sulphate, which, in turn leads to a reduction in the amount of ettringite formed. In the case of pulverized fuel ash, the resistance to sulphate attack is further improved because the presence of SO_3 in the ash means that the C_3A has already reacted with the sulphate, and thus is not available for further sulphate attack (Cabrera, 1986).

Despite the fact that the pozzolanas react with the $Ca(OH)_2$ in the concrete, there is little evidence that the pH of the concrete is significantly altered by their presence. It is stated that even very high concentrations of pulverized fuel ash do not reduce the pH of the concrete to the critically low values that could cause depassivation of the steel (Cabrera, 1986).

13.1.4 Microsilica additions

The effects of microsilica additions on the carbonation, chloride penetration and rate of corrosion of reinforcement is surveyed in a report by the FIP Commission on Concrete (FIP Commission on Concrete, 1988). They conclude that the incorporation of microsilica will delay the initiation of corrosion by chloride-induced corrosion. Microsilica may, however, shorten the initiation time for carbonation-induced corrosion, but only in low-to-medium grade concrete. Carbonation is not generally regarded as a problem in concrete with compressive strengths greater than 40 MPa. It is also pointed out that in practice the situation is more complex, since it is the combination of these factors which govern the risk of corrosion, and these are difficult to predict for different mixes. But the evidence does suggest that additions of microsilica which improve the durability of the concrete will also result in an improvement of the protection of embedded steel from corrosion.

Recently, concretes containing microsilica have been used in conjunction with calcium nitrite additions to protect steel reinforcement in concrete from corrosion (Berke *et al.*, 1988). In this case, the microsilica additions reduce the permeability of the concrete and the calcium nitrite is a pitting corrosion inhibitor which acts by stabilizing the passivating layer on the surface of the steel reinforcement.

13.1.5 Durability of high-performance concrete

One of the newer potential problems concerning the durability of high-performance concretes which incorporate microsilica in the mix is that of attack by magnesium sulphate. The microsilica reacts with the calcium hydroxide to form **CSH** gels by the pozzolanic activity of the microsilica, and should increase resistance of the concrete to attack by sulphates. In general this seems to be the case (FIP Commission on Concrete, 1988), but attack by magnesium sulphate on the **CSH** gel in these high-strength concretes could be serious. In studies on pure cement pastes made from Type I and Type V Portland cements, with and without microsilica additions, it was found that the addition of microsilica to the cements resulted in an increase of attack on the paste by magnesium sulphate (Cohen and Bentur, 1988). In the case of sodium sulphate, the reduced permeability of the microsilica modified paste resulted in a paste resistant to sulphate attack. More research is needed on the effects of magnesium sulphate on microsilica modified concrete, rather than on pure cement/microsilica pastes.

13.1.6 Measurement of durability

(a) Use of strength as a measure of durability

In the past the strength of concrete has been used as a measure of its durability because, for ordinary concrete, its strength was determined by the presence of pores within the concrete. Interconnected pores normally were the result of the use of high water/cement ratios and these led to a decrease of strength of the concrete.

The use of strength as a measure of durability has its drawbacks which are outlined in a paper by Pomeroy (1986) who questions the use of the 28-day strength for this purpose. Recent developments in both the materials used in concrete and the methods used to mix it have led to a steady increase in the 28-day strength of the set concrete due, in part, to the use of faster-setting cements. In the past, it could be assumed that the 28-day strength of the concrete was only a fraction of the ultimate strength attained after much longer times. The increase in the 28-day strengths that have been attained recently means that this is not always the case, and thus can lead to an overestimation of the potential durability of the concrete.

The use of the 28-day strength specification as a means of specifying the durability of concrete now presents many potential hazards. Any reduction in the amount of cement used in the mix leads to the possibility of later problems with corrosion etc. If the 28-day criterion is the only one used to assess the concrete quality, then less of the faster hydrating cement can be used with the same water additions while still meeting the strength criterion. This results in an increase in the water/cement ratio compared with what was used previously to meet the specified strength and since the cost of cement is high, there is an obvious temptation to use this higher ratio. The increased water/cement ratio produces a concrete which contains more open-pore porosity and hence is more prone to deterioration. Also, the higher heat of hydration of these cements can produce temperature gradients in the concrete which can, in turn, lead to thermal cracking.

(b) Long-term testing

Long-term tests for durability are normally carried out in the field. The results of such tests have been used to investigate the deterioration of concrete by sea-water, sulphates etc. (Regourd, 1983). The major drawbacks are that there are always a large number of uncontrollable variables which have to be taken into account in the analysis of the test results. Also the long time taken for these tests makes such field testing impractical for the immediate evaluation of new materials under development.

(c) Accelerated testing

Accelerated testing is normally carried out in the laboratory for the rapid evaluation of new mixes and materials in concrete. In these tests, the experimental conditions can be closely controlled and the number of variables limited. The increase in testing rate is normally achieved by increasing the temperature at which the tests are carried out or by increasing the concentration of corrosive or aggressive media (Regourd, 1983). The application of the results of such accelerated tests carried out under controlled conditions, should be carefully considered because deterioration of concrete is often the result of a sequence of events and not due to a single cause. Other problems encountered by increasing the temperature is that it will affect the rate of hydration of the cement components as well as the reactions under study. The increased curing rate of the concrete under test could possibly alter the rate of attack by aggressive media.

13.2 PROTECTION

13.2.1 Prevention of corrosion of steel in concrete

The underlying causes of the corrosion of steel in concrete are poor design, poor workmanship, and the use of calcium chloride to accelerate the set of the concrete. If there is an adequate cover of high-quality concrete over reinforcement, then its corrosion should not be a problem.

As mentioned previously, many methods have been used in attempting to prevent the corrosion of reinforcement in concrete. These include galvanizing the rebars in which the zinc coating will protect the steel from corrosion by a sacrificial action. If galvanizing is used, the finished product must be washed with a chromate solution to prevent the reaction of pure zinc with concrete to produce hydrogen. Stainless steel rebars have been used in critical areas and corrosion resistant steel containing nickel is reported to be under development in Japan (Gerwick, 1988), but, in both of these cases, extreme care has to be used to ensure that these special steels do not come into electrical contact with other reinforcement or metals in the concrete as corrosion due to the contact of dissimilar metals can then occur. Coating the rebars with epoxy resins effectively puts a high electrical resistance in the corrosion circuit and thus reduces corrosion, but this is only effective if the coating is undamaged. In critical applications, such as the prevention of corrosion of prestressed steel, cathodic protection techniques can be used. This involves the maintenance of the steel in the passivated state by the use of an impressed DC voltage. Practical methods by which this can be achieved are described by Berkeley and Pthmanaban (1990).

The methods described above directly protect the reinforcement from corrosion. Other methods to prevent corrosion by restricting the penetration of chloride ions, carbon dioxide, water etc. have been described earlier in this section.

Thin continuous film Thick dense layer

(a) (b)

Hydrophobic layer Pores blocked

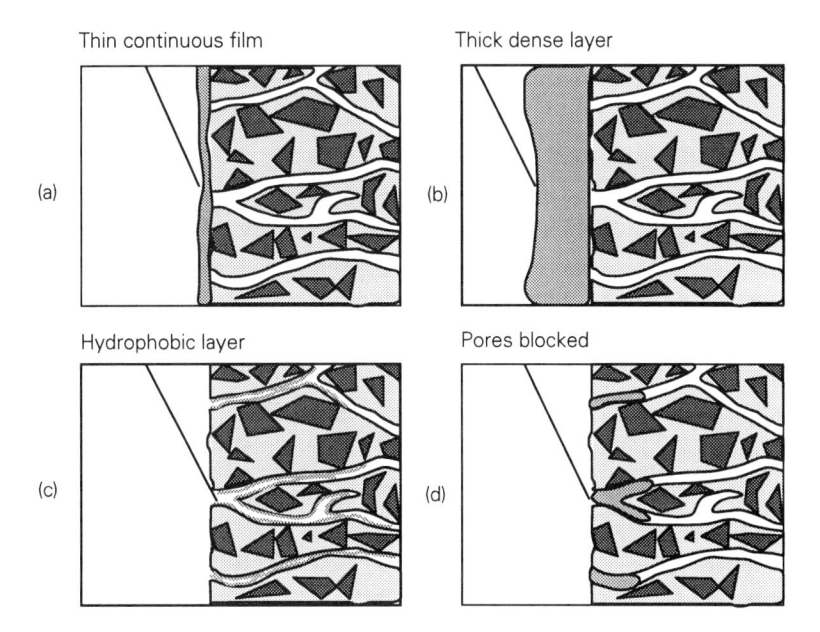

(c) (d)

Figure 13.1 Types of surface treatment: (a) sealer/coating, (b) rendering, (c) making pore surfaces hydrophobic and (d) blocking pores (redrawn from Keer, 1992)

13.2.2 Surface protective treatments

Surface protective treatments all have the effect of blocking the surface pores in the concrete and reducing permeability, but in most cases the depth of penetration is very small, and these treatments are only worthwhile when the situation is only mildly aggressive to concrete.

There are four main ways in which the surface of concrete can be sealed, and the choice of the method used will depend on the nature of the attack on the concrete. The methods of sealing the pores are shown in Figure 13.1 and described below.

These surface treatments are:

(a) Sealers/coatings

These must be impermeable to water, water vapour, oxygen and carbon dioxide. Suitable results can be obtained by painting the concrete with oils or synthetic resins, or coating with epoxy resins. Repeat coatings may be needed every few years.

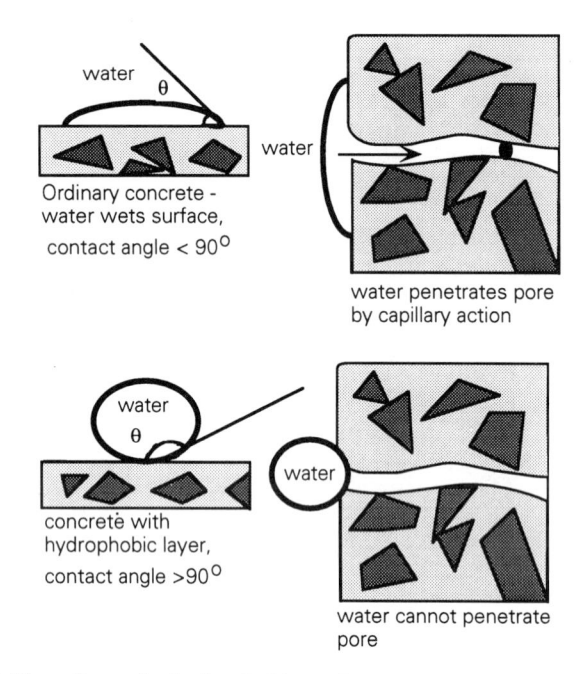

Figure 13.2 The effect of a hydrophobic surface on the penetration of water into pores

(b) Rendering

Rendering the concrete surface is done by the application of a thick layer of cement-based mortar. The mortar may be polymer-modified.

(c) Making pore surfaces hydrophobic

Hydrophobic surfaces can be obtained by treatment of the concrete with silicone based materials. These materials react with moisture to form silicone resins which are chemically bonded to the surface of the cement.

The hydrophobic surface prevents penetration of the water into the pores due to the high contact angle of the water on the surface. This is shown in Figure 13.2.

There are two major classes of silicon based materials that are used for hydrophobic surface treatment of concrete (McGettigan, 1992), these are the silanes, siloxanes and siliconates; and the silicones. The structures are shown in Figure 13.3.

The siliconates, siloxanes and silanes contain one organofunctional group and bond to the substrate as shown in Figure 13.4. The larger the organofunctional group, the higher the water repellency of the coating.

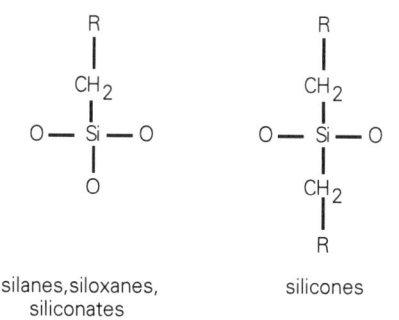

Figure 13.3 The structures of silanes, siloxanes and siliconates; and silicones

Methyl or ethyl groups have less water repellency than the higher molecular weight iso-butyl or n-octyl groups. Also a branched structure is more water-repellent than a linear chain. Thus the branched isobutyl group has about the same effect as the linear n-octyl group, which is about double the size of the isobutyl group.

The silicones have two organofunctional groups and are provided as polymers. They provide a hydrophobic surface, but are not bonded to the surface of the concrete, as shown in the diagram in Figure 13.5.

(d) Blocking pores

The chemical reaction of magnesium (or zinc) fluorosilicate with cement is an example of the pore blocking treatment. These react with the calcium compounds in the cement to form insoluble fluorides and the pores are blocked by the silicic acid or alumina hydrates which are formed during the reaction.

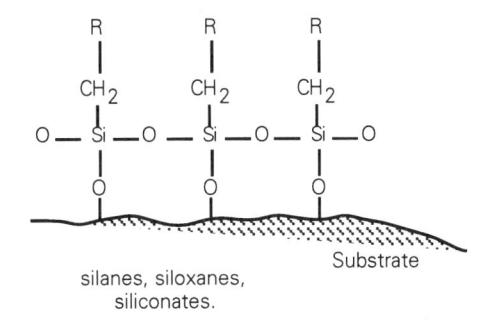

Figure 13.4 The bonding of the siliconates, siloxanes and silanes onto a substrate

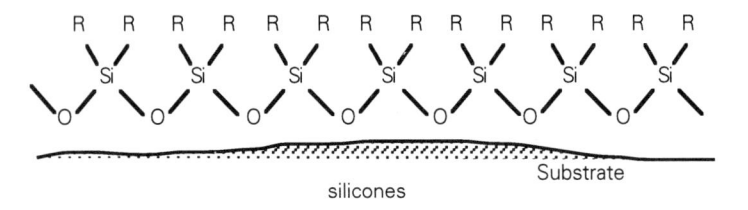

Figure 13.5 Silicones on a concrete surface

Sodium or ethyl silicate can also be used to block the pores with precipitated colloidal silica. The water-soluble sodium silicate is activated by contact with carbon dioxide and the alcohol soluble ethyl silicate can be either acid or base catalysed to be hydrolysed with water. The overall reaction can be written as:

$$Na_4SiO_4 + 2CO_2 \rightarrow SiO_2 + 2Na_2CO_3$$

or

$$Si(OC_2H_5)_4 + 2H_2O \rightarrow SiO_2 + 4C_2H_5OH$$

Impregnating with bitumen under pressure also results in the blocking of pores in concrete.

QUESTIONS

1. Discuss the effects of permeability and microcracks on the durability of concrete.
2. How do additions of pozzolanic materials affect the durability of concrete?
3. What is the relationship (if any) between strength and durability of concrete? Can the strength of concrete be used as a measure of durability?
4. What is accelerated testing? What are the advantages and disadvantages of accelerated testing compared with field tests?
5. How do surface treatments protect concrete from attack?

References

Berke, N. S., Pfeifer, D. W. and Weil, T. G. (1988) Protection against chloride-induced corrosion. *Concrete International*, **10**, 45–55.

Berkeley, K. G. C. and Pthmanaban, S. (1990) *Cathodic Protection of Reinforcement Steel in Concrete*, Butterworths, London; Boston.

REFERENCES

Cabrera, J. C. (1986) The use of pulverized fuel ash to produce durable concrete. In *Improvement of Concrete Durability*, Proceedings of the seminar 'How to make today's concrete durable for tomorrow', the Institution of Civil Engineers, Thomas Telford, London, pp. 29–57.

Cohen, M. D. and Bentur, A. (1988) Durability of Portland cement-silica fume pastes in magnesium sulphate and sodium sulphate solutions. *ACI Materials Journal*, **85**, 148–57.

FIP Commission on Concrete, (1988) *Condensed Silica Fume in Concrete*, Thomas Telford, London.

Gerwick, B. C. Jr. (1988) Concrete marine structures. In D. F. Hanson and C. R. Crowe (eds), *Materials for Marine Systems and Structures*, Academic Press, Boston, pp. 352–87.

Keer, K. G. (1992) Surface treatments. In G. Mays (ed.), *Durability of Concrete Structures*, E. & F. N. Spon, London; New York, pp. 146–65.

McGettigan, E. (1992) Silicon-based weatherproofing materials. *Concrete International*, **14** (6), 52–56.

Pomeroy, C. D. (1986) Requirements for durable concrete. In *Improvement of Concrete Durability*, Proceedings of the seminar 'How to make today's concrete durable for tomorrow', the Institution of Civil Engineers, Thomas Telford, London, pp. 1–27.

Regourd, M. (1983) Durability. Physio-chemical and biological processes related to concrete. In S. Rostam (ed.), *Durability of Concrete Structures*, Department of Structural Engineering, Technical University of Denmark, Lyngby, pp. 49–71.

Schießl, P. (1984) Protection of reinforcement. In S. Rostam (ed.), *Durability of Concrete Structures*, Department of Structural Engineering, Technical University of Denmark, Lyngby, pp. 241–97.

14 | Resistance of concrete to fire

14.1 FIRE DAMAGE TO CONCRETE

Concrete is widely used in buildings and could be exposed to the effects of fire at some time. If this happens it is necessary to be able to assess the amount of damage done to a structure by a fire in order to determine whether or not the structure can continue to be used. Of all the materials of construction used in a building, concrete is probably the least affected by fire and is often used to protect load-bearing steel beams from high temperatures. However it is accepted that the exposure of concrete to high temperatures will adversely affect its properties. If the concrete reaches a temperature of 250–300 °C it is generally recognized that a significant loss of strength starts to occur.

The depth of concrete that is exposed to temperatures in excess of this critical range will depend on a large number of factors which include the intensity and duration of the fire as well as the properties of the concrete itself. Even the most fire-resistant concretes will fail completely if exposed for a considerable period to temperatures exceeding 900 °C, while there will be serious reductions in strength when a temperature of 600 °C is attained.

The types of failure include cracking and spalling of surface layers, reduction of strength of the concrete, exposure of reinforcement to heat and loss of tension in prestressed tendons.

The fire resistance of concrete is primarily controlled by the type of cement used, the type of aggregate, and the thermal conductivity and heat capacity of the concrete. These factors determine whether the effects of high temperature persisting for a short time will be restricted to the surface material or whether they will be transmitted to the interior of the concrete. In the case of reinforced concrete, cold-worked steel rebars suffer from a reduction in strength if heated to above 450 °C whereas a reduction of strength of hot rolled steel commences at 600 °C. The effect of fire on prestressed high tensile steel tendons is critical and a reduction in properties can occur if the tendons are heated to above 200 °C (Mallett,

FIRE DAMAGE TO CONCRETE

1994). The fire resistance of reinforced concrete is dependent on the depth of concrete cover over the reinforcement.

The failure of concrete due to fire is mainly a result of the differential expansion of the hot surface layers and the cooler concrete in the interior. There is also the opposing actions of the cement, which normally shows a net shrinkage and the aggregate, which expands with increasing temperature. These effects lead to the cracking and spalling of the concrete that may be severe enough to expose reinforcement. Exposed reinforcement conducts heat rapidly and the resulting expansion of the reinforcement relative to the surrounding concrete causes further damage.

14.1.1 The effect of temperature on the cement

Fully cured Portland cement will first expand on heating, but this expansion is opposed by shrinkage which tends to become dominant at higher temperatures. If these movements are not uniform throughout the cement, cracking of the cement will occur.

Both physical and chemical changes occur when cement paste is heated. Between 20 and 110 °C there is a loss of evaporable water. This can cause internal vapour pressure build-up if the water vapour cannot escape rapidly enough from the surface. At temperatures between 250 and 300 °C the hydrates in the cement decompose. At higher temperatures decomposition of calcium hydroxide and calcium carbonate then takes place and it is virtually complete by 700 °C (Illston *et al.*, 1979).

The production of calcium oxide (quick lime) by decomposition of calcium hydroxide or calcium carbonate presents a potential problem which manifests itself after the cement has cooled. If the cooled concrete comes into contact with water the calcium oxide will rehydrate and expand. This effect could cause disruption of the concrete which has withstood a fire without actual disintegration. It has been found that this problem is less in concrete that has had additions of blast furnace slag. This is thought to be due to the removal of the calcium hydroxide by the slag during the setting reactions.

14.1.3 The effect of temperature on the aggregate

Aggregates that have been subjected to high temperature in their formation or manufacture show the best resistance to fire. The most resistant of all concretes to fire are those made with blast furnace slag aggregate, although broken brick aggregate also shows good fire resistance providing it is free of quartz. The least resistant aggregate materials might be thought to be those which decompose as the temperature increases (such as limestone) but in this case the decomposition of dense material does not occur until 900 °C and actual fire tests have shown that, except under conditions of prolonged exposure, only decomposition of the surface of the aggregate takes place. There is some debate as to whether quartz aggregate causes problems since at high temperatures it undergoes sudden expansion at 573 °C due to the α to β phase change. Orchard (1979) suggests that the effect of this expansion might be absorbed by the 'plasticity' of the cement paste, but Lea (1970) considers that the expansion would be catastrophic and that concretes with aggregates of siliceous gravels, flint and granite show very poor fire resistance.

Assessment of the performance of aggregates in fires can be gained by measuring their thermal expansion or contraction as a function of temperature by dilatometry and the temperatures at which physical or chemical changes take place can be determined by differential thermal analysis and differential thermogravimetric analysis.

14.1.3 The effect of thermal conductivity of concrete

Concrete damage due to cracking and spalling will be minimized if the thermal conductivity of the concrete is low because this will restrict rapid temperature rise, and therefore damage, to the surface layers of the concrete. The thermal conductivity of concrete can be reduced by the use of a lightweight porous aggregate, and/or by the entrainment of air in the concrete. Lightweight aggregates are used in the manufacture of fire-resistant lightweight concrete, but this is at the expense of its strength.

The effect of thermal conductivity is complex, because the thermal conductivity of concrete is a function of temperature. Generally the thermal conductivity decreases with increasing temperature due to the increase in porosity of the concrete caused by the evaporation of pore water and dehydration of the cement paste. In a fire the surface of the concrete will undergo these reactions and this will produce a porous, heat insulating layer at the surface. This low thermal conductivity layer will then reduce the rate at which the temperature rises in the bulk of the concrete.

14.1.4 The effect of the heat capacity of concrete

In general, the higher the heat capacity of the concrete, the higher will be the short-term resistance to fire. This is because the temperature rise in concrete with high heat capacity will be less than that which takes place in a concrete with a low heat capacity for the same exposure conditions. The heat capacity of concrete is mainly controlled by the heat capacity of the aggregate materials used in the concrete.

14.1.5 The effect of the presence of microsilica in the concrete

With the increasing use of high-performance concrete (incorporating microsilica and superplasticizers) there was concern about how this new material would behave in a fire. One report was published that stated that concrete containing microsilica exploded when exposed to high temperatures. However, as was pointed out by Shirley *et al.* (1988), these tests had been carried out on a mortar (not concrete) containing more than 20 wt% microsilica and having a very high compressive strength of 172 MPa. Fire tests that were carried out on high-strength concrete slab samples (with and without microsilica) and a normal strength concrete slab showed that there was no significant difference between the fire endurance of the specimens. No explosive behaviour was observed, and none of the samples showed spalling on the exposed surface. Some cracking over embedded reinforcement did occur but was observed in both the silica containing and silica free samples. Sanjayan and Stocks (1993) compared the behaviour of two full sized T-beams which had been prepared in identical manner and dried indoors before testing. The slabs were prepared so that the effects of different thickness of cover over reinforcement and different slab thicknesses could be assessed. It was found that the 105 MPa HPC exhibited explosive spalling when exposed to fire, whereas the 27 MPa NSC withstood the test. The spalling was confined to the region on the slab which had the thickest coverage of the steel, and it was concluded that residual moisture in this section had a significant influence on the spalling. The presence of the moisture in the thick section of the HPC could be linked with the slower drying rate of the HPC when compared with the more porous NSC.

A description of other tests carried out in order to determine the fire resistance of high-performance concrete is given in a report on microsilica in concrete by the FIP Commission on Concrete (1988). It was concluded that none of the studies carried out indicated that increased strength led to increased damage on exposure to fire. One study did show that damage was done to 50–60 MPa concrete exposed to a hydrocarbon fire if the prestressed or loaded concrete was in a moist state, but the damage was not dependent on the mix composition; the microsilica content simply increases the likelihood of moisture being retained in the concrete.

14.1.6 The effect of fire on steel-fibre-reinforced concrete

The behaviour of steel-fibre-reinforced concrete beams exposed to fire has been studied by Kamal *et al.* (1992). Their preliminary experiments showed that after exposure to fire, the steel-fibre-reinforced beams had higher flexural rigidity, higher initial cracking loads and withstood higher ultimate loads than similar reinforced beams that had been cast without fibres; presumably this could be due to the increased net thermal conductivity of the steel-fibre-reinforced concrete. The morphology of the fibres also had an effect. Triangular twisted rough surfaced fibres showed better performance than plain round fibres. This was attributed to the better

Table 14.1 Classification of fire damage to concrete (Chung, 1994)

Intensity of fire	Surface of concrete	Surface colour	Crazing	Spalling	Exposure of reinforcement
Light	Some peeling	Unchanged	Slight	Slight	None
Moderate	Substantial loss	Pink	Noticeable	Localized	10–25%
Severe	Total loss	Buff	Extensive	Extensive	24–50%
Very severe	Total loss	Buff	Extensive	Extensive	>50%

mechanical and thermal bond between the triangular fibres and the concrete paste.

14.2 ASSESSMENT OF FIRE-DAMAGED CONCRETE

In order to determine the extent of damage done to a concrete construction after a fire (so that decisions can be made whether to repair or rebuild) it is necessary to have an estimate of the temperature to which the concrete has been exposed. The temperatures attained in a fire may sometimes be estimated by colour changes that the cement or aggregates undergo after exposure to heat, but not all concrete formulations behave in the same way and their colour changes are strongly dependent on the iron content of the concrete. A classification of fire damage to concrete is shown in Table 14.1.

Hardness measurements can also be used, but the inhomogeneous nature of concrete means that the assessment is subject to error of both measurement and interpretation.

Differential thermal analysis (DTA) is a standard technique commonly used to determine the temperature at which various physical and chemical changes take place when a substance in heated. Most of the physical and chemical changes that occur on heating concrete are irreversible; once the concrete has been heated to a certain temperature the changes that took place on the initial heating will not be observed when the concrete is reheated in a DTA test. Thus comparison of the DTA trace obtained from a fire-exposed sample of concrete with the trace obtained from a non-exposed sample can give an indication of the temperature to which the concrete had been previously exposed.

Riley (1991) has developed a method that uses the petrographical examination of thin sections of concrete in order to assess the temperature isotherms within the concrete, so that the 300 °C isotherm can be located.

He found that concrete exposed to temperatures above 500 °C became anisotropic to plane polarized transmitted light. At temperatures between 300 and 500 °C cracks were formed within the cement paste (mortar) and around aggregate particle boundaries, whereas below 300 °C there was only boundary cracking around aggregate particles. Thus the onset of intrapaste cracking can be used to locate the 300 °C isotherm. The study, however, was only carried out on concrete containing one type of aggregate and therefore is of limited use at the present.

It has been proposed by Chew (1988) that thermoluminescence could be used to assess the temperature of exposure of the concrete. Thermoluminescence is the production of light by some materials that occurs when these materials are heated. Measurement of its thermoluminescence spectrum as a function of temperature can give an indication of the thermal history of the concrete as well as the duration of exposure to that temperature. In a later paper, Chew (1993) describes the effects of long exposure times on the thermoluminescence, and finds that for temperatures above 230 °C, the effects are not linear and the estimated length of exposure can be in error.

QUESTIONS

1. What type of concrete would you use to protect load-bearing steel beams from the effects of fire? Give reasons for you answer.
2. Describe the changes that take place when concrete made from Portland cement is exposed to fire.
3. How would you measure the fire resistance of aggregate materials?
4. Why is it essential to have adequate concrete coverage of tendons in prestressed concrete for fire resistance?
5. What methods can be used to determine the amount of damage done to a concrete structure after exposure to fire?

References

Chew, M. Y. L. (1988) Assessing heated concrete and masonry with thermoluminescence. *ACI Materials Journal*, **85**, 537–43.

Chew, M. Y. L. (1993) Effect of heat exposure duration on the thermoluminescence of concrete. *ACI Materials Journal*, **90**, 319–22.

Chung, H. W. (1994) Assessment of damages in reinforced concrete structures, *Concrete International*, **16** (3), 55–9.

FIP Commission on Concrete, (1988), *Condensed Silica Fume in Concrete*, Thomas Telford, London.

Illston, J. M., Dinwoodie, J. M. and Smith, A. A. (1979) *Concrete, Timber and Metals*, Van Nostrand Reinhold, New York, p. 556.

Kamal, M. M., Bahnasawy, H. H. and El-Refai, G. (1992) Behaviour of fibre reinforced concrete beams exposed to fire. In R. N. Swamy (ed.), *Fibre Reinforced Cement and Concrete*, E. & F. N. Spon, London, New York, pp. 764–74.

Lea, F. M. (1970) *The Chemistry of Cement and Concrete*, 3rd edn, Edward Arnold, London, Ch. 19.

Mallett, G. P. (1994) *Repair of Concrete Bridges*, Thomas Telford, New York, p. 8.

Orchard, D. F. (1979) *Concrete Technology*, 4th edn, Vol. 1, Applied Science Publishers, London, Ch. 6.

Riley, M. A. (1991) Assessing fire-damaged concrete, *Concrete International*, **13** (6), 60–3.

Sanjayan, G. and Stocks, L. J. (1993) Spalling of high-strength silica fume concrete in fire. *ACI Materials Journal*, **90**, 170–3.

Shirley, S. T., Burg, R. G. and Fiorato, A. E. (1988) Fire endurance of high-strength concrete slabs. *ACI Materials Journal*, **85**, 102–8.

Special cements and concretes | 15

Many special cements and concretes have been developed for a wide range of applications. Most are more expensive than ordinary Portland cement. Some examples of these materials are given in this section. The list is by no means comprehensive and only an outline will be given of the production and properties of these materials. The systems to be described include:

- high alumina cement used as a construction material in concrete;
- high alumina cement used as a refractory castable;
- fast-setting cements based on high alumina cement;
- polymer-modified cements and concretes and high-strength polymer mortars (Macro Defect Free) based on Portland cement or high alumina cement;
- supersulphated cement which is resistant to sulphate attack;
- modifications to Portland cement based materials; these include the production of coloured cement and concrete by the control of impurities in the raw materials and by the use of pigments; concrete used for nuclear radiation shields produced by the additions of special aggregates and cements and concretes which expand during the setting reactions;
- non-calcareous cements for refractory applications, for use in acidic or salt environments and for dental applications.

15.1 HIGH ALUMINA CEMENT

High alumina cement (HAC) was specially developed in 1908 by Jules Bied in France to resist sulphate attack of concrete linings of railway tunnels which ran through rocks that contained high concentrations of magnesium and sodium sulphates. It had been known since the mid nineteenth century that some calcium aluminate compounds could be used as cements, but it was not until the problems of sulphate attack on Portland cements became severe, that alumina-based cements were produced for commercial use.

Table 15.1 Colour and composition ranges (mass%) of high alumina cements

Type	Colour	Al_2O_3	Fe_2O_3	SiO_2	CaO
1	Grey-black	37–40	11–17	3–8	36–40
2	Light grey	48–51	1–1.5	5–8	39–42
3	Cream-grey	51–60	1–2.5	3–6	30–40
4	White	72–80	0–0.5	0–0.5	17–27

HAC is resistant to sulphate attack due to the absence of **C₃A** (tricalcium aluminate) and **CH** (calcium hydroxide) which are both involved in the attack on Portland cement by sulphates.

The raw materials used to make HAC are limestone ($CaCO_3$) and bauxite ($Al(OH)_3$). The bauxite is usually contaminated with other minerals which can contain Fe_2O_3, TiO_2 and SiO_2. The presence of iron or titanium oxidesdo not adversely affect the HAC, but the silica content can be a problem and must be kept low. The colour of the HAC is dependent on the iron oxide content of the bauxite.

Four main types of HAC are produced, the compositions of which are shown in Table 15.1.

The main difference between the four types is the alumina and iron content, type 4 being virtually iron free. The major use of type 4 is for the production of high-temperature castable refractory.

In the production of types 1, 2 and 3 HAC, the limestone and bauxite are completely melted (as distinct from the partial melting of reactants used in the production of Portland Cement clinker). This complete melting is the origin of the trade name Ciment Fondu (melted cement). It is also known in other countries by other trade names such as Cemento Fuso, or Schmelzzement (Robson, 1962, p. 10).

Various types of kilns have been used to produce these high alumina cements, ranging from blast furnaces (in which metallic iron can be produced as well as HAC), reverberatory furnaces, electric arc furnaces, and rotary kilns developed in the USA similar to those used to produce Portland cement clinker. The main difference between the rotary kiln used to make HAC and that used for the production of Portland cement clinker is that the region of the kiln which contains the molten HAC (at 1500–1600 °C) is not refractory lined but employs water cooling of the steel shell so that a layer of frozen HAC is maintained to protect the shell (Robson, 1962, p. 22). The molten HAC flows out through holes in the rotary furnace and is water cooled to produce granules which are then ground to a fine powder.

Type 4 HAC is produced by either solid state or liquid phase sintering of the lime and bauxite. In this case there is either no melting or at the best only partial melting of the components.

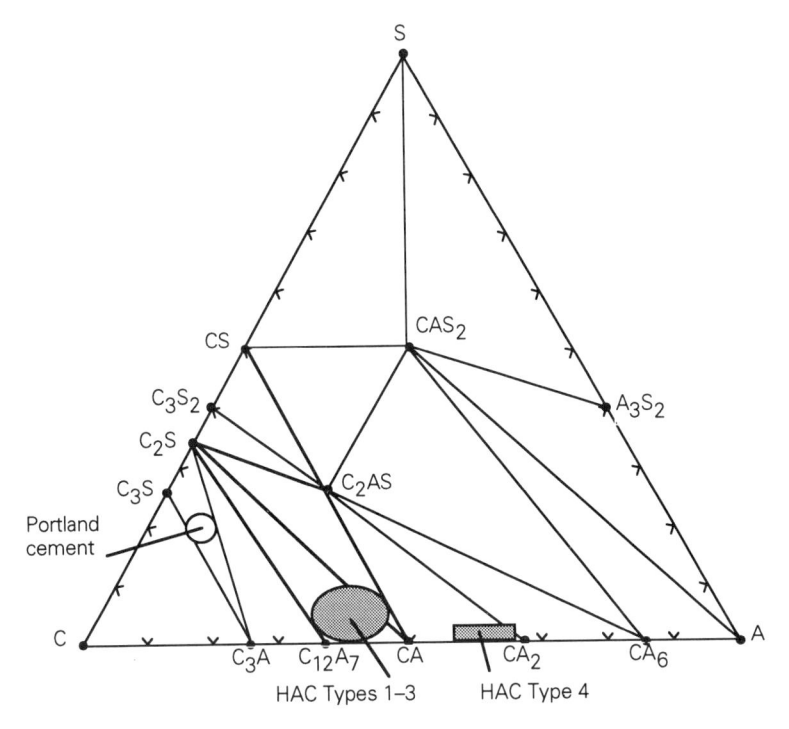

Figure 15.1 Compositions of high alumina cement (mole%)

15.1.1 Composition of the HAC clinker

The range of compositions of the HAC clinker are shown in the calcia-alumina-silica phase diagram in Figure 15.1.

The main compound that is formed in types 1 to 3 HAC is **CA** (mono-calcium aluminate). The other components are **C₂S** and either **C₁₂A₇** or **C₂AS**. Titania (TiO₂ = T) and iron oxide impurities in the bauxite may also lead to the formation of **CT** and **C₂F-C₂A** solid solution (**C₄AF**).

The strength of the HAC is primarily due to the hydration of **CA**. The ordinary aluminous cements lie entirely in the field of stability of **CA**, as shown in Figure 15.1, so it is this phase that will first crystallize out of the liquid HAC after it is tapped from the furnace. The cooling rate of the HAC is important as it affects the microstructure of the clinker. Slow cooling of the liquid results in the maximum separation of **CA** and produces a cement with high hydraulic activity. Rapid cooling can lead to the production of clinker containing 5–35% glassy phase which produces cements with slower setting and hardening rates. However, the final strength of the cement is unaffected by cooling rate of the molten HAC (Robson, 1964).

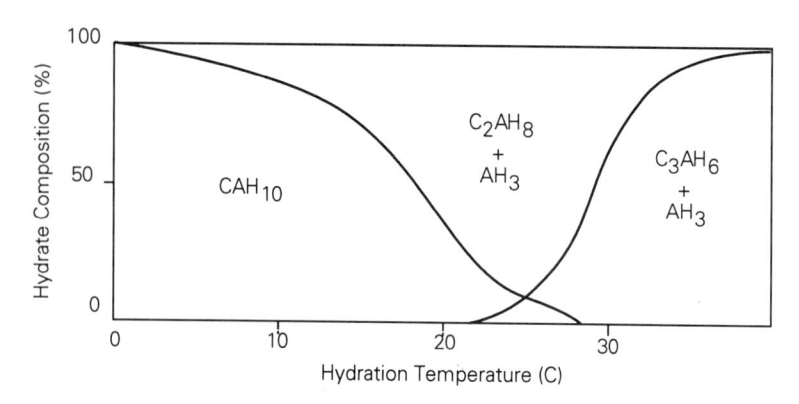

Figure 15.2 Hydration products of HAC (redrawn from George, 1983)

The fact that the production of these cements involves complete melting of the components means that equilibrium can be achieved and the final phases present in the clinker can be reasonably well predicted from the equilibrium phase diagram (although rapid quenching of the clinker will favour glass formation).

It is more difficult to use the equilibrium phase diagram to predict the phases present in type 4 refractory cement (white HAC) since there is either no melting or only partial melting of the components and equilibrium is often not achieved. White HAC consists of **CA** as either the sole cementitious agent (the other phase being alumina), or as the main agent (the other phases being alumina and **CA$_2$**).

15.1.2 Hydration reactions of HAC

The products of hydration are dependent on temperature as shown in Figure 15.2.

At low temperatures, (<15 °C), the major hydration reaction of **CA** is:

CA + 10H → CAH$_{10}$

This produces hexagonal **CAH$_{10}$**, with a small quantity of the other hydrated products. Between 15 °C and ~30 °C, the major reaction is:

2CA + 11H → C$_2$AH$_8$ + AH$_3$

This produces hexagonal **C$_2$AH$_8$** and amorphous **AH$_3$** together with smaller quantities of other hydrates.

In the same temperature range, other HAC components react:

C$_{12}$A$_7$ + 51H → 6C$_2$AH$_8$ + AH$_3$

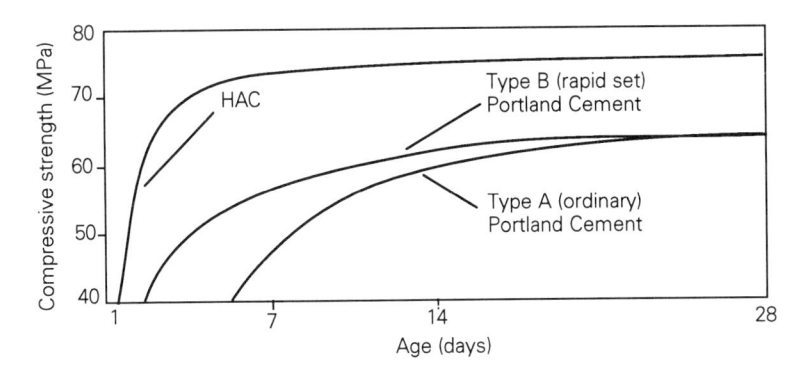

Figure 15.3 Rate of development of strength of HAC, as compared with ordinary Portland cement and rapid setting Portland cement

and

$$C_2S \rightarrow CSH_{(gel)}$$

At these low reaction temperatures (<30 °C) the CAH_{10} or C_2AH_8 form interlocking platelets and these give the cement its strength.

Above 30 °C cubic C_3AH_6 and AH_3 are the main hydration products.

The amount of water which combines chemically with the anhydrous cement to produce the hydrates is about 50% of the mass of the cement, about twice as much as required for the hydration of Portland cement. As a consequence, for the same water/cement ratio in a mix, the HAC has a lower porosity than an equivalent Portland cement mix. Also the rounded nature of the cement grains of the HAC, compared with those of Portland cement serves to improve the workability of HAC cement (Neville, 1975, p. 18).

15.1.3 Uses of HAC for buildings

The increase in strength of the cement is very rapid, as shown in Figure 15.3.

The initial set of HAC takes place from two to six hours after mixing, and the final set not more than two hours later. The gain in strength is such that within 24 hours the compressive strength can reach 90% of the ultimate strength of the concrete.

The rapid attainment of strength has obvious advantages when this cement is used to make concrete for structural purposes. It means that the formwork can be removed quickly and thus the building process is shortened. There are also obvious advantages in using HAC in prestressed concrete as the rapid strength gain means that the overall time taken to produce pre- or post-tensioned structural components can be significantly reduced.

However, the rapid hydration of the cement is accompanied by a rapid evolution of its heat of hydration and care must be taken to prevent excessive temperature rises in the setting concrete. Wet curing is strongly recommended for the first 24 hours after pouring. This not only ensures that the cement has sufficient water for hydration, but also is a most effective means of removal of the heat generated by the hydrating cement (Neville, 1975, p. 19).

Despite its high strength, the use of HAC for structural purposes has now fallen into disrepute, due to many sudden failures. These occurred mainly in situations where concrete made from HAC was exposed to relatively high temperatures and high humidity, for example, the failure of beams supporting the roof over an indoor swimming pool at a school in England in 1974 (Neville, 1975, Ch. 7). Investigations of the failed beams revealed that the structure of the HAC had changed. It was found that the CAH_{10} and C_2AH_8 crystals that were primarily responsible for the strength of the HAC were actually metastable crystalline phases, and with time they transformed to the stable phase C_3AH_6:

$3CAH_{10} \rightarrow C_3AH_6 + 2AH_3 + 18H$ (1/2 the original volume)

$3C_2AH_8 \rightarrow 2C_3AH_6 + AH_3 + 12H$ (2/3 the original volume)

The C_3AH_6 has a cubic structure and forms rounded crystals, and these do not contribute to the strength of the cement that was due to the interlocking platelets in the original cement. There is also an increase in porosity due to the C_3AH_6 being more dense that the CAH_{10} and C_2AH_8. The rate of these phase changes is dependent on temperature and humidity. The effect of temperature is shown in Table 15.2.

The failure of the swimming pool roof was not unique, the history of failures dates back to the mid 1930s, but in most cases the correct cause was not identified, and the blame was laid on incorrect cement manufacture, incorrect water/cement ratios, the effect of sodium chloride, attack by sea water, poor workmanship etc. An interesting account of failures of high alumina cement and concretes is given by Neville (1975).

Table 15.2 The effect of temperature on the time for half conversion to C_3AH_6

Temperature (°C)	Time for half conversion to C_3AH_6
50	1 week
40	100 days
25	20 years

15.1.4 Use of HAC as a refractory castable cement

Portland cement starts to lose strength at about 300 °C and between 400–600 °C the Ca(OH)$_2$ dehydrates. It can only withstand 900 °C for very short times. HAC is very satisfactory as a refractory castable material (white HAC, type 4). A reduction of strength occurs when the freshly cast material is first heated up to 900 °C, but above this temperature strong ceramic bonds are formed and the cement can be used up to 1800 °C. The maximum temperature of use of the castable is dependent on the type of aggregate that is used in the castable. For lower temperature uses, prefired clay (chammotte or grog) is used, but for higher temperatures sillimanite (Al$_2$O$_3$.SiO$_2$), mullite (3Al$_2$O$_3$.2SiO$_2$) or fused alumina are used as aggregates.

15.2 FAST-SETTING AND HARDENING CEMENTS (KURDOWSKI AND SORRENTINO, 1983)

Fast-setting and hardening cements are all based on modified HAC, in which the composition has been adjusted so that no **CA** is formed. The compositions of these fast-setting cements vary, but most contain **C$_{12}$A$_7$**, **C$_2$S** and minor amounts of one of **C$_4$A$_3$$\bar{\text{S}}$**, CaF$_2$ or CaCl$_2$. Some free lime and MgO is also present. It appears that it is the **C$_{12}$A$_7$** that is critical for the fast set. The setting time is adjusted by partial substitution of the modified HAC by Portland cement.

As well as the chemical composition affecting the setting time of these cements, other physical factors can be adjusted to give the required setting time. These include the fineness of the cement and the temperature of the mix. Increasing the fineness and the temperature (to about 30 °C) will both increase the rate at which the cements set. In most cases there is a compromise that is made between the rate of setting and the ultimate strength of the cement.

15.2.1 Hyper-fast setting

Hyper-fast-setting cements consist of pure modified HAC and have setting times of less than three minutes. These cements can be used for instantaneous space filling, for example the control of water movement in sewers and caves, or the caulking of boats, and for immediate mechanical support.

15.2.2 Very-fast setting

The setting time of these cements is between three and eight minutes. Very-fast-setting and fast-setting cements are often made from admixtures of the hyper-fast set aluminous cement with ordinary Portland cement.

The cements can be used for drying up moist walls and tanks, cement rendering in water and in building stairs, sinks, washstands etc.

15.2.3 Fast-setting

Fast-setting cements set in 8 to 60 minutes. They can be used for the installation of thin dividing walls, sealing of inspection covers and manholes, door and window frames and ledges.

The composition of these fast-setting cements is similar to those for the very-fast-setting cements but with a larger proportion of the Portland cement in the mixture.

15.3 POLYMER-MODIFIED CEMENTS (MORTARS) AND CONCRETES

Polymer additions are made to mortars and concretes in order to increase such properties as tensile or flexural strength, toughness, adhesion and resistance to chemical attack.

Polymer-modified mortars and concretes may be classified into three categories (Chandra and Ohama, 1994, p. 81):

- Polymer-modified mortar and concrete in which polymer additions are made to mortar or concrete mixes based on inorganic cements (e.g. Portland cement, HAC etc.).
- Polymer mortar and concrete where the polymer acts as the phase which binds the aggregate particles together. There are no inorganic cements present in the mix.
- Polymer-impregnated mortar and concrete. These are conventional inorganic-based mortars and concretes are impregnated with polymers after they have set. This process is expensive and is normally only used for repair work when other methods cannot be used.

Only the first group of polymer-modified mortars and concretes will be considered here.

15.3.1 Production

Polymer-modified mortars and concretes are produced by the partial replacement of the inorganic cementing phases present in ordinary mortars and concrete by polymers. Conventional mixing methods are used to produce the mortars and concretes. The wide range of polymers that can be added can be divided into four classes: aqueous latex suspensions, polymer powders that are redispersed in water, water-soluble polymers and liquid polymers (Chandra and Ohama, 1994, p. 89). The amount of

POLYMER-MODIFIED CEMENTS (MORTARS) AND CONCRETES

polymer added to the mix varies and is dependent on the ultimate use of the mix. Cement/polymer ratios (by weight) of up to 1:1 can be used, but 1 : 0.5 is more typical (Chandra and Ohama, 1994, p. 91).

(a) Aqueous latex suspensions

The addition of latex particles suspended in water results in these particles coating the hydrating cement particles. This leads to the formation of an interpenetrating polymer/cement hydrate phase (e.g. **CSH**) which serves to bind together the aggregate particles in the mortar or concrete.

(b) Redispersed polymer powders

Redispersible polymer powders (e.g. ethylene vinyl acetate) act in a similar manner to the latex additions. The major difference is that the polymer powders are normally added to the dry mortar or concrete mix and are dispersed in the water during the concrete mixing process (Chandra and Ohama, 1994, p. 86).

(c) Water-soluble polymers

Water-soluble polymers (e.g. polyvinyl alcohol, PVA) are added as powders to the wet mix during the mixing of the mortar.

(d) Liquid polymer

Modification with liquid resins which are added during mixing (such as epoxy resins or unsaturated polyesters). Polymerization of the resins by water takes place as the cement hydrates and results in the formation of an interpenetrating network of polymer and cement hydrates. These form a strong bond between the sand and aggregate particles and results in an increase of strength of the mixture.

15.3.2 Effects of polymer additions on mortars and concretes

(a) Before setting

Polymer additions increase the workability of the mix. This is thought to be due to the effect of the polymer in reducing the friction between the solid aggregate particles and also to the effect of entrained air in the mix (Chandra and Ohama, 1994, p. 111). Air entrained in the mix is stabilized as small air bubbles by the action of the polymers. The amount of entrained air can be excessive in some cases and anti-foaming agents then have to be added. In most polymer-modified mortar mixes, high strength is often not needed and the amount of entrained air is between 5 and 20 volume %, whereas in polymer-modified concretes it is limited to about 2 volume %, similar to that of unmodified concrete.

The slump for a given water/cement ratio is increased by polymer additions, which means that the water/cement ratio can be decreased. The use of a lower water/cement ratio results in an increase in strength and a decrease in drying shrinkage of the modified mix. As mentioned previously, water retention of the wet mix is improved. Bleeding and segregation is also prevented in the modified systems.

(b) Setting behaviour

The coating of all components of the modified mortar or concrete by polymers results in a delay in setting. The amount of retardation of set is dependent on the polymer used (Chandra and Ohama, 1994, p. 112).

The complex interaction between the polymer and cement particles during hydration will obviously be dependent on the type of polymer that is used. However, in general, the polymers influence the nucleation and growth of the calcium hydroxide crystals (**CH**) formed during the hydration of the **C$_3$S** (Chandra and Ohama, 1994, p. 160). The polymer coating formed around the **CH** crystals inhibits their growth and this results in the formation of a large number of small **CH** crystals instead of a few large crystals. The adsorption of the polymer chains on the **CH** also results in the surface of the coated crystals being hydrophobic instead of hydrophilic. The polymers also retard the formation and growth of ettringite formed by the reaction of gypsum with **C$_3$A** and this also result in the formation of many small ettringite needles (Chandra and Ohama, 1994, p. 164).

(c) Properties and uses of polymer-modified mortars and concretes

In general, polymer-modified mortars and concretes have increased flexural and tensile strengths, when compared with the unmodified materials, but the compressive strength remains unchanged for a mix with the same water/cement ratio. The actual change in flexural and tensile strength is dependent on the type of polymer and the polymer/cement ratio. Similar to normal mixes, the curing conditions and test conditions also have an effect (Chandra and Ohama, 1994, p. 113). The increase in tensile and flexural strength means that the impact and wear resistance of these materials are greatly increased and this has led to the modified concretes being used for flooring, pavements and airport runways.

Other properties, such as shrinkage, creep and thermal expansion, are dependent on the polymer used and can be larger or smaller than for the unmodified mortars or concretes (Chandra and Ohama, 1994, p. 118).

One of the advantages of the addition of polymers is the effect that they have on reducing the water absorption, water permeability and water vapour penetration into the set concrete (PVA is an exception, although the film formation that occurs during the setting process does reduce water loss from the mortars). The lower open-pore porosity of the modified concretes is also reflected in increased resistance to carbonation, chloride ion and oxygen penetration and hence reduced corrosion of reinforcement. This means that the concretes can be used as anticorrosive linings of effluent drains, chemical plant floors, septic tanks etc. There have, however, been some corrosion problems when polymers containing chlorine (e.g. poly (vinylidene chloride-vinyl chloride)) have been used. These have been shown to release chloride ions which have themselves caused corrosion problems of the reinforcement.

Another advantage is the good adhesion that polymer-modified mixes have to other materials (cement mortar or concrete, stone, metals, wood, tiles etc). This is attributed to the high adhesion of polymer to these materials. Adhesives for floor and wall tiles are normally made of polymer-modified materials. The good adhesive properties mean that the mortars are successfully used as repair materials for cracks and delaminations in existing damaged concrete etc.

As would be expected, the fire resistance of the modified materials is dependent on the nature and amount of polymer present, but for low polymer/cement ratios of up to about 15 % (by wt) the material is rated as incombustible (Chandra and Ohama, 1994, p. 129).

The entrained air, which is characteristic of these materials, means that the freeze/thaw resistance is good. The polymers have a reproducible pore structure as well as a good dispersion of air in the modified mortar. However, it has also been shown that polymer additions that are non-air entraining also result in increased freeze/thaw resistance. It is thought that this comes about by the effect of the polymers on the size of the **CH** crystals which reduces the microcrack formation during the crystallization process.

15.3.3 Macrodefect free cement (MDF cement)

Macrodefect free cements are also produced from cement/polymer mixtures. The major difference from the polymer-modified cements described in the previous section arises from the method of production of these mixtures. The methods of mixing used in the production of MDF have been adapted from those used in the processing of polymers. In the production of the MDF cements, a water-soluble polymer, such as polyvinylacetate, is mixed with either Portland or high alumina cement and a limited amount of water is added (Ford, 1990). This polymer/cement dough is then mixed and air removed from the dough under reduced

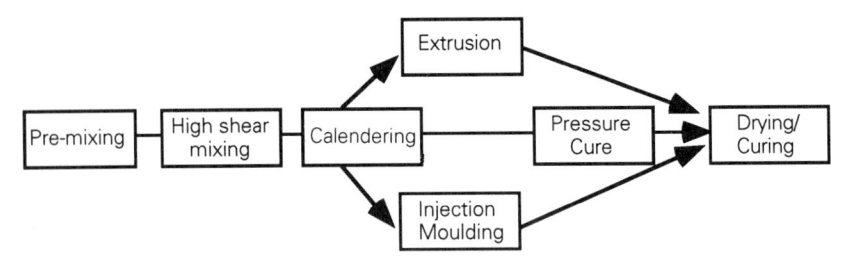

Figure 15.4 MDF cement process flow chart (redrawn from Ford, 1990)

pressure. The dough is then processed using conventional plastics technology, for example, by extrusion. During curing, the polymer chains are initially adsorbed onto the cement grains and, during hydration, the hydration products intermingle with the polymer chains. As the water is removed by the hydration of the cement, the polymer chains dehydrate and contract. The whole body shrinks, and there are no macrodefects left in the system. Only a limited degree of hydration of the cement occurs. The fabrication route is shown in the flow chart of Figure 15.4.

The absence of macrodefects results in a dramatic increase in the strength of the cement. Ordinary Portland cement which contains flaws of ~1 mm in size has a 3-point bending strength of ~20 MPa and a Young's modulus of 20 GPa, whereas the MDF cement, in which the pore size is reduced to a few micrometres, exhibits a bending strength of ~150 MPa (comparable to aluminium) and a modulus of up to 45 GPa (Birchall *et al.*, 1982). The fracture toughness of MDF cements is similar to that of grey cast iron, but this can be greatly increased by the incorporation of fibre reinforcement.

Many variations of MDF cement compositions have been produced. The cement can either be Portland cement, high alumina cement, calcium phosphate cement etc. Different water-soluble polymers can be used, and these can be combined with various fillers and fibres.

Applications for MDF cements that have been investigated include materials for acoustic damping (cement loudspeakers) and ceramic armour (Alford and Birchall, 1985).

The use of these cements as substrates for printed circuit boards is under investigation. They are less sensitive to moisture than fibre-reinforced epoxy resins, and it is expected that they will be competitive with alumina substrates. Incorporation of iron powder and nylon fibres into the MDF cement results in a mix that could have use for shielding sensitive electromagnetic equipment from radiated or conducted electromagnetic interference.

15.4 SUPERSULPHATED CEMENT

Supersulphated (slag) cement is made largely from blast furnace slag. It is produced from a mixture of 80 to 85 mass% slag, 10 to 15 mass% dead burnt gypsum (anhydrous calcium sulphate) and 5 mass% Portland cement clinker. The components are ground together more finely than is normal for Portland cement. The ground cement has a very low heat of hydration and is very sensitive to moisture and therefore must be stored under very dry conditions before use.

Supersulphated cement's advantages lie in its resistance to sea-water and sulphate attack. The sulphate resistance of this cement is second only to that of HAC. It also shows resistance to acid (peaty) water and to oils. Concrete made from supersulphated cement is resistant to acidic solutions down to pH 3.5, providing the water/cement ratio used in the concrete mix is below 0.45 (Neville, 1981, p. 76).

Supersulphated cements have been used mainly in Europe, but their use is limited. This is in spite of the fact that they are primarily made from materials which can be considered as waste products (blast furnace slag from iron making and gypsum from processes such as flue gas desulphurization). According to Bijen and Niël (1982), the main reason for this limited use is that carbonation of the surface of the set cement results in dusting of the surface, and there can also be an increase in porosity of the surface and corresponding decrease in strength due to carbonation. However, Neville (1981, p. 393) points out that this loss of strength due to carbonation of the surface of the cement is not structurally significant. One of the other problems associated with these cements is that variations in the composition of the blast furnace cements that are used to make the cements result in variations in the properties of the cements. Bijen and Niël (1982) describe research into the effects of variations of chemical composition of the slag, and the use of additives to improve the properties of the cements.

The reactions that occur on hydration of the supersulphated cements are complex. Ettringite is formed in the early stages of hydration and is responsible for the initial hardening and strength development (El-Hemaly et al., 1987). At a later stage **CSH** is formed and this contributes to the strength of the paste. The overall reaction, as given by Tashiro and Okubo (1982), can be simplified as:

slag (**S + C + A**) + gypsum (**C$\bar{\text{S}}$H$_2$**) + slaked lime (**CH**) + water (**H**)

$\rightarrow$ ettringite (**C$_6$A$\bar{\text{S}}$H$_{32}$**) + CSH (**C$_5$S$_6$H$_5$**) + calcium aluminate gel (**C + A + H**)

They stated that the strength development at ordinary temperatures was due firstly to the dissolution of alumina from the surface of the slag (controlled mainly by the concentration of **C**). This is followed by the formation of ettringite and **CSH** (controlled by the **C** and **S** concentrations) and finally the production of other hydrates. The use of high-pressure steam curing to attempt to increase the hydration rate was found to alter the reaction products and results in a weaker, more porous structure. This is in contrast to the increase in strength which can be achieved by steam curing of Portland cement-based materials.

Supersulphated cement hardens at a slower rate at low temperatures compared with Portland cement.

15.5 MODIFICATION OF PORTLAND CEMENT-BASED MATERIALS

15.5.1 White cement and concrete

Portland cement normally has a grey colour due to the presence of iron oxide. In order to produce a white cement, impurities in the raw materials used in the production of the cement must be kept to a very low level. To produce white cement, china clay, which contains little or no iron or manganese oxides, is fired with iron-free chalk or limestone. The fuel used in the production of the cement clinker is normally gas or oil, since the use of coal results in the addition of coal ash to the clinker and this coal ash can be a source of iron impurities. Special precautions also have to be taken in the grinding of the clinker to avoid contamination. This all increases the cost of the cement to about double that of Portland cement. Because of the cost increase, the white cement is often used only as a coating over ordinary concrete (Neville and Brooks, 1987, p. 30).

15.5.2 Coloured cement and concrete

The surface of concrete can be coloured by the application of paint, but a more durable and low-maintenance colour can be produced by the incorporation of 5–10% coloured pigments into the concrete (Lynsdale and Cabrera, 1989). Ideally a pigment should be a fine dry powder, or an aqueous suspension or slurry of the powder, inert to the ingredients of concrete. In practice, pigments may be inorganic or organic materials. In order to obtain a uniform colour, the pigment must have a fine particle size (<1 μm). The pigment used must be stable to the effects of light and weather, and not adversely affect the setting properties of the cement. They must also be alkali stable. Pigments that fulfil these conditions are shown in Table 15.3.

The differences in colour obtained by using iron oxides are due firstly to the different colours of the various oxides (yellow, black and red), and

Table 15.3 Pigments used to produce coloured cement and concrete

Pigment	Colour
Iron oxides	Red, yellow, brown, black
Manganese oxide	Black, brown
Chromium oxide	Green
Cobalt blue	Blue
Copper phthalocyanine	Blue, green
Ultramarine blue	Blue
Carbon	Black

secondly to the effect of particle size and shape of any one oxide. For example, iron oxide red pigments can vary in colour from violet red to brick red depending on their particle size and shape which varies from cubical to spherical. Brown iron oxide colour is obtained by mixing black, yellow and/or red.

The level of dosing of the concrete with the pigment affects the intensity of the colour. The maximum amount of pigment that can be added, according to American and British standards, is 10% of the mass of cement, however the normal range used is from 3% to 6%. The intensity of the colour is also dependent on the water/cement ratio used, (the lower the ratio, the more intense the colour).

In most cases the incorporation of pigments result in a reduction in the workability of the concrete, and the inclusion of wetting agents and surface active agents is necessary when hydrophobic pigments like carbon black and copper phthalocyanine are used in order to improve their dispersion and to reduce the water needed. Other methods of overcoming this problem are to make the entire mix hydrophobic by the addition of water repellent agents (e.g. stearic acid) or to mix the pigment with fine silica in order to make it hydrophilic.

Care has to be taken that the pigments do not contain harmful fillers, which are sometimes added to reduce the cost of the pigment. Common fillers used in pigments are calcium carbonate (chalk), barium sulphate and gypsum. Chalk and barium sulphate are harmless but the presence of large proportions of gypsum can result in sulphate attack on the cement.

Pigments may either increase or lower the strength of the cement, the reduction of strength being the greatest with the addition of carbon black. Similarly the effect on shrinkage is variable. Carbon black has also been reported to accelerate the setting of concrete and even produce a flash set (Lynsdale and Cabrera, 1989).

15.5.3 Concretes for use as radiation shields

Concretes that are used for shielding x-ray and γ-ray equipment (e.g. that used for non-destructive testing) and for nuclear radiation shielding are made by the addition of barium sulphate aggregates to the cement paste. The additions result in an increase in the density of the concrete, and this increases the absorption of the radiation. Lead oxides cannot be used, since they adversely affect the setting reactions of the cement. Other high-density aggregates that can be used include magnetite, ilmenite ($FeTiO_3$), and limonite ($2Fe_2O_3.3H_2O$). Metallic iron or steel shot can also be added (Lea, 1970).

15.5.4 Expansive (or expanding) cements

According to Neville (1994), expansive cements were developed in Russia and in France in the 1940s. The cements consisted of Portland cement mixed with an expanding agent and a stabilizer. The expanding agent was a mixture of calcium sulphate and calcium aluminate and the stabilizer was blast furnace slag, which was used to react with the excess calcium sulphate and bring the expansive reactions to an end.

Expansive cements can be used to counteract the drying shrinkage of Portland cement and also, when used in reinforced concrete, to prestress the concrete. In the latter case the concrete expands during hardening, but this expansion is constrained by steel reinforcement, which results in the steel being put into tension and the concrete into compression (Neville and Brooks, 1987, p. 32).

When using these cements in concrete it is essential that the time over which the cement expands is controlled normally to between four and seven days. If the time of expansion is too long, then the cement will have hardened to such an extent that the continuing production of expansive material will result in cracking of the hardened concrete. If the expansive cement is to be used to prestress the concrete then the expansion should not take place in less than four days as the strength of the concrete mix might not be sufficiently well developed and could still accommodate the expansive strain without putting the reinforcement into tension. The latter is inefficient and costly since the expansive ingredient is expensive (Cohen and Mobasher, 1991). Control of the expansion time can be achieved by careful control of the cement particle size and particle size distribution as well as the chemical composition of the mix. The finer the cement grains, the faster the expansion (Cohen and Richards, 1982). Additions of microsilica to the mix has been shown to result in an increase of the rate of expansion (Lobo and Cohen, 1992) and this helps to minimize the damaging effect of excessive expansion after setting (Cohen et al., 1991). Most of these cements owe their expansive properties to the production of ettringite which causes the cement paste to expand as it sets. The ettringite is formed by the reaction of compounds with gypsum and lime. It should be noted that the formation of ettringite requires larger amounts of water than in the setting of Portland cement so that it is essential to wet cure expansive cements (Neville, 1994).

(a) Types of expansive cements

(i) Type M
Type M (ASTM C 845-80) cement is made by grinding together Portland cement clinker, high alumina cement clinker and gypsum. The resultant cement is quick setting, rapid hardening and, when set, resistant to sulphate attack. The expansion is produced within two to three days after casting. This cement was first marketed in 1970.

The expansion occurs by the reaction of **C_3A** (from the Portland cement) and **CA** (from the HAC) with lime, gypsum and water to form ettringite. **$C_{12}A_7$** in the HAC will form ettringite, but its presence is normally considered undesirable because its very rapid reaction with water makes the setting process difficult to control. This property is exploited in the production of the rapid-setting cements.

The reactions that cause the expansion are:

$C_3A + 3C\bar{S} + 32H \rightarrow C_6A\bar{S}_3H_{32}$ (ettringite)

$3CA + 3C\bar{S} + 32H \rightarrow C_6A\bar{S}_3H_{32} + AH_3$ (ettringite and aluminium hydroxide)

In the presence of calcium hydroxide, the aluminium hydroxide transforms to **CAH_{10}** (Kurdowski and Sorrentino, 1983):

$CH + AH_3 + 6H \rightarrow CAH_{10}$

The **CAH_{10}** can then react with gypsum to form ettringite (in the presence of **CH**):

$CAH_{10} + 2CH + 3C\bar{S} + 20H \rightarrow C_6A\bar{S}_3H_{32}$ (ettringite)

(ii) Type K
Type K (ASTM C 845-80) cement is made by adding an expanding agent to Portland cement and blast furnace slag. The expanding agent is formed by heating a mixture of gypsum, bauxite and chalk to form $4CaO.3Al_2O_3.SO_3$ **($C_4A_3\bar{S}$)**. This is often called the Klein complex, after Alexander Klein, who first proposed its use and patented it in 1961. The use of this phase as the expansive agent is convenient as it hydrates very quickly and the process is normally finished after seven days (Kurdowski and Sorrentino, 1983). When the cement is mixed with water, the excess

calcium sulphate is taken up by the blast furnace slag and this reaction controls the expansion of the cement. The rate and magnitude of the expansion is controlled by careful proportioning and mixing of the ingredients.

The expansive reactions are:

$$C_4A_3\bar{S} + 8C\bar{S} + 6C + 74H \rightarrow 3C_6A\bar{S}_3H_{32} \quad \text{(ettringite)}$$

$$C_3A + 3C\bar{S} + 32H \rightarrow C_6A\bar{S}_3H_{32} \quad \text{(ettringite)}$$

(iii) Type S

Type S cement is also specified in the same standard. It has a high C_3A content and contains slightly more calcium sulphate than is normal in Portland cement. The expansive reactions are similar to those in Type K. Type S was introduced in 1968 by Greening from the Portland Cement Association.

(iv) Type O

Type O was introduced in Japan in 1965 (Cohen *et al.*, 1991). This expands by the action of calcium oxide, which has to be specially processed.

(b) Theories on the mechanism of expansion

The details of how the expansion occurs is still a matter of debate. Cohen (1983) describes the two main theories that have been advanced to explain the mechanism of expansion, namely, a crystal growth theory and a swelling theory.

In the crystal growth theory, it is stated that the expansion is a direct result of the crystallization pressure which is due to the formation and growth of ettringite crystals which are formed on the surfaces of the reacting particles or in the solution. The theory assumes that as soon as the reactions commence, the surfaces of the expansive particles are covered by a dense coating of ettringite. This coating increases in thickness due to further hydration, and when the coating thickness increases to above that of the surrounding solution, the adjacent particles are pushed apart, and expansion occurs. This expansion will continue until either the dissolved gypsum ($C\bar{S}$) or the expansive particles ($C_4A_3\bar{S}$, C_3A) are used up.

The swelling theory assumes that the expansion is caused by an ettringite gel which swells by water absorption.

The points of agreement between the two theories are as follows.

- The expansion is affected by the size of the ettringite crystals and the presence of lime.
- Ettringite is formed as small crystals which contribute to expansion in the presence of lime, sulphates, water and expansive particles.

- If lime is absent, then the reaction of the sulphates and water result in the production of large ettringite crystals which do not contribute to the expansion.

The points of disagreement are as follows.

- The crystal growth theory says that hydration is favoured and more rapid in the presence of lime, and that the ettringite is formed on the surface of the expansive phases. Only in the absence of lime would the ettringite form in locations away from the surface of the expansive particles.
- The swelling theory says that the ettringite always forms by a through-solution mechanism, regardless of the absence or presence of lime, and that the rate of hydration of the expansive particles is smaller in the presence of lime (Cohen, 1983).

Kurdowski and Sorrentino (1983) identify four hypotheses that have been put forward to explain the expansion. These are:

- the pressure of crystallization due to the anisotropic growth of the crystals;
- the *in situ* crystallization of the anhydrous phases of the hydrated product (the crystal growth theory);
- the absorption of water by the colloidal ettringite (the swelling theory); and
- expansion by osmotic pressure.

(c) Uses of expansive cements and concretes

The type K expansive cement is the most commonly used, since both the rate and amount of the expansion are said to be more reliably predictable than those of the other formulations (Neville, 1981, p. 86).

The first major use is to compensate for drying shrinkage of cements and concretes. About 10 to 15% of the expansive component is added to the mix in order to obtain a relatively low unrestrained expansion of about 0.2%.

The second major use is to produce self-stressing concretes and cements. These contain a higher amount of the expansive component (25–40%) in order to obtain a high potential expansion. During the setting of the concrete, steel reinforcement in the concrete restrains the expansion in the concrete, and as a result it is put in tension, and this produces compressive stress in the set concrete. The volume change that occurs in self-stressing cement, shrinkage compensating cement and ordinary Portland cement is shown in Figure 15.5.

The small residual expansion of the shrinkage compensating cement is aimed at leaving the concrete in enough compression only to prevent the development of shrinkage cracks (Neville, 1994).

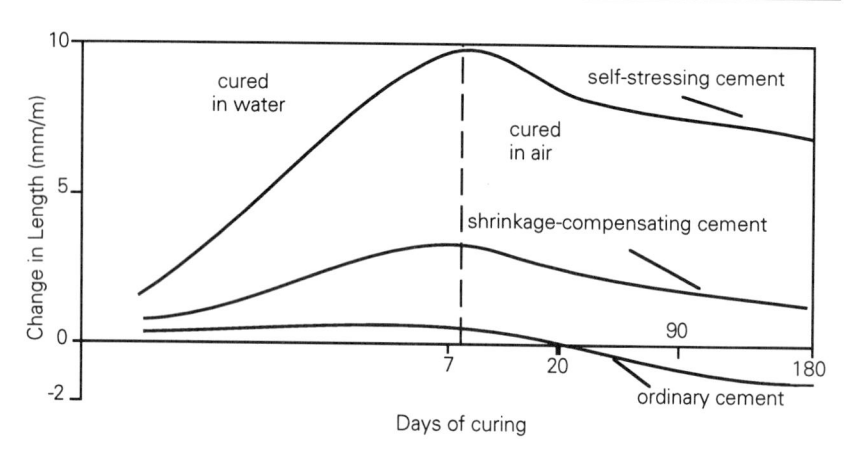

Figure 15.5 Comparison of volume changes of cements (redrawn from Kurdowski and Sorrentino, 1983)

These cements still have only limited use mainly because great care must be taken in their use. It is essential that a homogeneous mix is obtained and it is necessary that the expansion be predictable both in extent and rate.

15.6 NON-CALCAREOUS CEMENTS

These cements have no calcium compounds in them. Portland cements contain both calcium silicates and calcium aluminates. Non-calcareous cements can be prepared by the substitution of calcium by strontium or barium.

15.6.1 Strontium and barium cements

Some of the strontium aluminates and barium aluminates have cementing properties when mixed with water, similarly the strontium and barium silicates will also contribute to the setting of these cements. Cements containing barium silicates have a high resistance to attack by sea water and sulphates due to the insoluble nature of the product of reaction, barium sulphate (Kurdowski and Sorrentino, 1983). Compounds that possess hydraulic setting properties include those shown in Table 15.4.

The main use of these cements, particularly those containing the aluminates, is as high-temperature refractory materials. Barium aluminate concrete can be used to higher temperatures than the corresponding HAC concrete.

NON-CALCAREOUS CEMENTS

Table 15.4 Strontium and barium hydraulic setting compounds

$3SrO.SiO_2$	(**Sr₃S**)
$SrO.Al_2O_3$	(**SrA**)
$SrO.2Al_2O_3$	(**SrA₂**)
$3SrO.Al_2O_3$	(**Sr₃A**)
$3BaO.Al_2O_3$	(**Ba₃A**)
$BaO.Al_2O_3$	(**BaA**)

Cements based on barium and strontium aluminates have been developed mainly in the USSR (Petzold and Rohrs, 1970, p. 41). $BaO.Al_2O_3$ (**BaA**), with a melting point of 1830 °C, has been recommended for fire resistant cements. On heating, the hydrates decompose and reach a minimum strength between 600 and 700 °C. This compares very favourably with the behaviour of Portland cement, as is shown in Figure 15.6 in which the cold compressive strengths after heating to various temperatures are graphed (redrawn from Petzold and Rohrs, 1970).

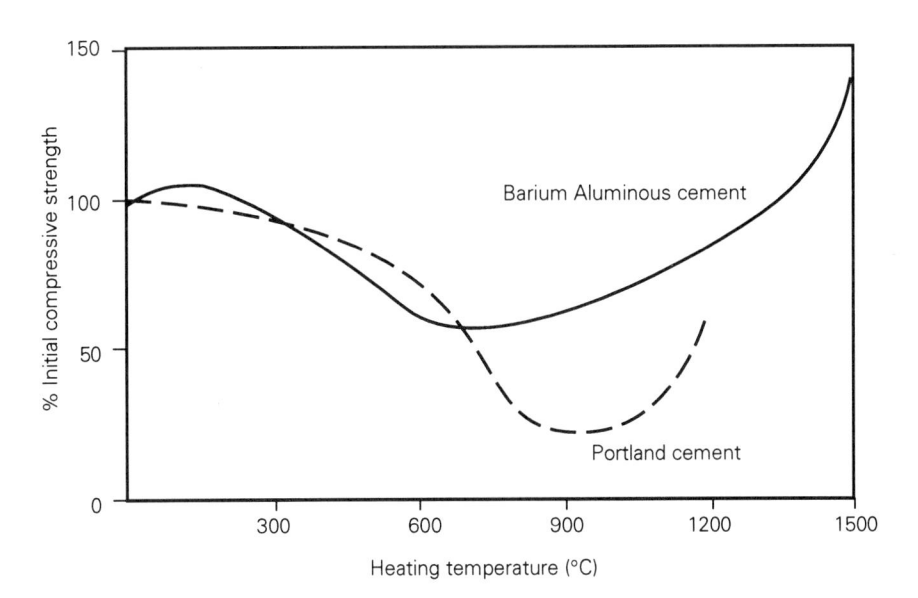

Figure 15.6 Cold compressive strength after heating the materials to various temperatures

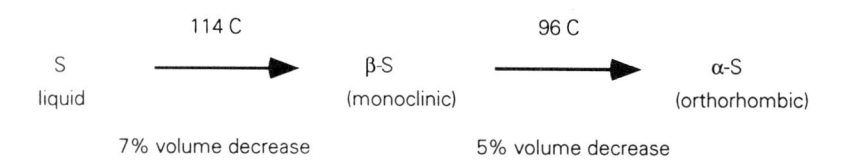

```
                114 C                    96 C
    S      ─────────────►    β-S    ─────────────►    α-S
  liquid                  (monoclinic)          (orthorhombic)

         7% volume decrease        5% volume decrease
```

Figure 15.7 Phase changes that occur when sulphur is cooled

The setting reactions are complex, and the presence of impurities in the cements can affect them, but the main ones are as shown below.

BaA + 6H → BaAH$_6$

2 (BaAH$_6$) → Ba$_2$AH$_9$ + AH$_3$

Ba$_2$AH$_9$ + BaA → Ba$_3$AH$_6$ + AH$_3$

It is the formation of the alumina trihydrate (**AH$_3$**) that gives the initial hardening of the cement, and the final set is due to both the formation of **AH$_3$** gels and **Ba$_3$AH$_6$**.

Barium aluminate cements have also been suggested for the use in nuclear reactors as radiation shields (Petzold and Rohrs, 1970, p. 213).

15.6.2 Sulphur concrete (ACI Committee 548, 1988)

Sulphur concretes are relatively new materials, and find use in areas exposed to acidic or salt environments. The thermoplastic sulphur concrete is prepared by mixing hot sulphur cement with mineral aggregates. Early attempts at making sulphur concrete by mixing sulphur with aggregate resulted in variable and unreliable mixes. This was, in part due to the various phase changes that occur when sulphur is cooled and the volume changes that accompany the phase changes, as shown in Figure 15.7.

The shrinkage which takes place when the solidified concrete is cooled through sulphur's β to α phase change results in the sulphur binder becoming highly stressed and prone to cracking.

This problem was overcome by the modification of the sulphur with additives such as olefinic hydrocarbon polymers or cyclopentadiene. The latter results in the stabilization of the β form of sulphur to low temperatures.

The appearance of sulphur concrete is similar to that of concrete prepared from Portland cement, but its strength (compressive, flexural and tensile) and fatigue life is superior to normal Portland cement concrete. Strength development is rapid, 70% of its ultimate strength is developed after a few hours. It is resistant to acids and salts, sets very rapidly and can be placed in below-freezing temperatures. It also has

Table 15.5 Properties of sulphur concretes

Cement (mass%)	Aggregate (mass%)	Voids (vol.%)	Compressive strength (MPa)	Workability
10.0	90.0	13.7	20.4	Relatively dry
12.5	87.5	9.7	42.1	Relatively dry
15.0	85.0	6.2	50.8	Stiff
17.5	82.5	5.5	51.4	Fluid
20.0	80.0	5.1	45.0	Soupy

very low permeability to water. There are, however, problems associated with handling of molten sulphur and suitable precautions have to be taken.

The aggregates used have to be selected according to the proposed use of the concrete. Silica (quartz) is stable in acid and salt environments, whereas limestone is only suitable for salt environments. Typical properties of sulphur concretes using <9 mm aggregate are shown in Table 15.5.

Steel reinforcement can be used, and glass fibres are very effective in control of shrinkage cracks and improving the toughness of the sulphur concrete.

15.6.3 Chemically bonded cements

Chemically bonded cements are based on mixtures of oxides (MgO, ZnO, glasses) with acids or salts which chemically react with the surface of the oxides to form chemical bonds between the particle. They have diverse applications, ranging from refractories to dental ceramics (Kurdowski and Sorrentino, 1983).

(a) Sorel cements

Sorel cements are based on magnesium oxide and magnesium chloride. Sorel cements have been known for many years, and concretes based on these cements have been widely used as industrial floors and as high-temperature refractories. The marble like appearance of the concrete has led to its use for decorative wall panels. The one major drawback of these cements is the lack of resistance to water, they are both water-soluble and release corrosive agents (Kurdowski and Sorrentino, 1983). The materials set by the reaction of magnesium oxide with magnesium chloride. The reaction products are complex, but the main products are $Mg_3(OH)_5Cl.4H_2O$ and $Mg_2(OH)_3Cl.4H_2O$. The water resistance of the cement can be improved by the reaction of these compounds with carbon dioxide from the atmosphere to form $Mg_2(OH)Cl(CO_3)3H_2O$ and

$Mg_5(OH)_2(CO_3)_4 \cdot 4H_2O$ which are insoluble in water. Other methods of improving the water resistance is by the addition of inorganic compounds such as mixtures of calcium sulphate or silicates or organic materials such as resins.

Since the discovery of the large magnesite deposits in Queensland, there has been renewed interest in their use as refractory materials.

(b) Dental cements

Dental cements are made by mixing a powder with either an acid or an alcohol. Their physical properties should be consistent with their use as fillings, crowns or bridges, namely suitable setting times, low heats of hydration, high volume stability, high strength and low porosity. The adhesion of the cement to the tooth in the presence of saliva must be good, and the cement should be resistant to dissolution and discoloration. They must also be non-toxic and biocompatible (Kurdowski and Sorrentino, 1983).

Materials that possess these properties include aluminosilicate with phosphoric acid, zinc oxide with phosphoric acid and zinc oxide with eugenol. Eugenol is an alcohol which reacts with the zinc oxide to form zinc eugenolate, which is used in 'temporary fillings' to protect the pulp of the tooth from the cement used for the temporary filling. Aluminosilicate cements are also used to fill teeth and are fast setting. They are made from sodium-calcium-fluoro-aluminosilicate mixed with phosphoric acid. The strength is similar to that of tooth enamel, as is the colour and thermal expansion coefficient. However there is a problem with erosion and disintegration in the oral environment. This has led to the replacement of the phosphoric acid with a solution of polyalkenoic acid. This acid can also be used in conjunction with zinc phosphate.

The aluminosilicate used is actually a glass material which is produced by the rapid quenching of a melt of suitable composition. The setting reaction between the glass and phosphoric acid is very complex and involves the leaching of the calcium and aluminium ions from the glass and subsequent precipitation of phosphates which serve to bind the glass particles together and set the mixture.

QUESTIONS

1. What is high alumina cement (HAC), and why was it developed?
2. What is the main difference in the production of HAC clinker (types 1–3) compared with the production of Portland cement clinker?
3. What is the main hydraulic component of HAC? Describe the effect of temperature on the hydration products of HAC.

4. Why is HAC resistant to sulphate attack?
5. Why was HAC used extensively for structural purposes in the past?
6. What was the cause of sudden catastrophic failures of HAC concrete and what conditions brought about the failures?
7. The composition of HAC has to be modified in order to produce hyper-fast-setting cements. Does this involve and increase or decrease of the alumina content of the HAC?
8. What is added to hyper-fast-setting cements to slow down the rate of reaction?
9. How are macro-defect free cements produced? Why do these materials have high flexural strengths?
10. What is supersulphated cement? Under what conditions would you use this cement?
11. In the production of coloured cements, why is the additions of pigment normally limited to under 10 mass%?
12. What is the important component in concrete mixes used for nuclear radiation shielding, and what properties should it have?
13. Describe the role of ettringite in the production of expansive cements and concretes. What is the effect of microsilica additions to expansive cement mixes?
14. What are the main uses of the expansive cements?
15. Why are cements based on barium silicate resistant to sulphate attack?
16. What is sulphur concrete?
17. What causes the sulphur concrete to set?
18. What changes occur when sulphur is cooled, and how do these affect the concrete?
19. What is Sorel cement?
20. Describe the desirable properties of inorganic dental cements.

References

ACI Committee 548 (1988) Guide for mixing and placing sulfur concrete in construction. *ACI Materials Journal*, **85**, 314–25.

Alford, N. McN and Birchall, J. D. (1985) The properties and potential applications of macro-defect-free cement. In J. F. Young (ed.), *Very High Strength Cement-Based Materials*, Materials Research Society, Pittsburgh, Pa. pp. 266–76.

Bijen, J. and Niël, E. (1982) Supersulphated cement: improved properties. *Silicates Industriels*, **48**, 45–53.

Birchall, J. D., Howard, A. J. and Kendall, K. (1982) A cement spring. *Journal of Materials Science Letters*, **1** (3), 125–6.

Chandra S. and Ohama Y. (1994) *Polymers in concrete*, CRC Press, Boca Raton.

Cohen, M. D. and Mobasher, B. (1991) Effects of sulphate and expansive clinker contents on expansion time of expansive-cement paste. *Cement and Concrete Research*, **21**, 147–57.

Cohen, M. D. and Richards, C. W. (1982) Effects of particle sizes of expansive clinker of type K expansive cements. *Cement and Concrete Research*, **12**, 717–25.

Cohen, M. D. (1983) Theories on expansion in sulfoaluminate – Type Expansive Cements: Schools of Thought. *Cement and Concrete Research*, **13**, 809–18.

Cohen, M. D., Olek, J. and Mather, B. (1991) Silica fume improves expansive-cement concrete. *Concrete International*, **13** (3), 31–7.

El-Hemaly, S. A. S., Galal, A. F., Mosalamy, F. H. *et al.* (1987) Activation of granulated blastfurnace slag by waste sulphated ash. *Silicates Industriels*, **52**, 9–12.

Ford, R. G. (1990) New developments in high performance cementitious materials. *Journal of the Australian Ceramic Society*, **26** (2), 163–70.

George, C. M. (1983) Industrial aluminous cements. In P. Barnes (ed.), *Structure and Performance of Cements*, Applied Science, London, pp. 415–70.

Kurdowski, W. and Sorrentino, F. (1983) Special cements. In P. Barnes (ed.), *Structure and Performance of Cements*, Applied Science, London. pp. 471–554.

Lea, F. M. (1970) *The Chemistry of Cement and Concrete*, 3rd edn, Edward Arnold, London, p. 539.

Lobo, C. and Cohen, M. D. (1992) Hydration of type K expansive cement paste and the effect of silica fume: I. Expansion and solid phase analysis. *Cement and Concrete Research*, **22**, 961–9.

Lynsdale, C. and Cabrera, J. (1989) Coloured concrete. *Concrete*, **23** (7), 29–34.

Lynsdale, C. J. and Cabrera, J. G. (1989) Coloured concrete: a state of the art review. *Concrete*, **23** (7), 29–34.

Neville, A. M. and Brooks, J. J. (1987) *Concrete Technology*, Longman Scientific & Technical, Harlow, Essex, UK.

Neville, A. M. (1975) *High Alumina Cement Concrete*, Construction Press, Lancaster.

Neville, A. M. (1981) *Properties of Concrete*, 3rd edn, Pitman, London.

Neville, A. M. (1994) Whither expansive cement? *Concrete International*, **16** (9), 34–5.

Petzold, A. and Rohrs, M. (1970) *Concrete for High Temperatures* (translated by A. B. Phillip and F. H. Turner), Maclaren & Sons, London.

Robson, T. D. (1962) *High-Alumina Cements and Concretes*, J Wiley & Sons, London.

Robson, T. D. (1964) Aluminous cement and refractory castables. In H. F. W. Taylor (ed.), *The Chemistry of Cements,* Vol 2, Academic Press, pp. 3–35.

Tashiro, C. and Okubo, Y. (1982) Characterization of supersulphate cement hardened under various curing, *Silicates Industriels*, **48**, 173–80.

Appendix A:
Australian standard Portland and blended cements

The classification used for cements in Australia has recently changed from prescription-based specifications to performance-based specification. The reasons for the change are given in amendment No. 2 to AS 3972 – 1991 which was published in August 1996 which 'state the characteristics desired by users without regard to the specific means to be employed in producing the product' and 'engineers/specifiers do not need to be concerned with the details of cement chemistry (e.g. constituents of the cement) as it is the final performance that is important. Therefore performance-based cement Standards are considered to benefit not only users, by leaving no doubt that their "engineering" requirements are met, but also producers, by allowing innovation and the development of new, improved materials and applications'.

The new classifications are

A.1 GENERAL PURPOSE CEMENT

Type GP – General purpose Portland cement, which contains Portland cement, calcium sulphate and up to 5% of mineral additions. The compressive strength at 7 days is 25MPa and at 28 days 40 MPa.

Type GB – General purpose blended cement, which is based on Portland cement with greater than 5% additions of fly ash or blast furnace slag, or both. The compressive strength at 7 days is 15 MPa and at 28 days 30 MPa.

A.2 SPECIAL PURPOSE CEMENTS

Type HE – High early strength cement. The compressive strength at 3 days is 20MPa and at 7 days 30 MPa.

Type LH – Low heat cement. The compressive strength at 7 days is 10 MPa and at 28 days 30 MPa and the heat of hydration a maximum of 280 J/g at 7 days and 320 J/g at 28 days.

Type SR – Sulphate resisting cement. The compressive strength at 7 days is 20 MPa and at 28 days 30 MPa and the maximum C_3A content 5 wt.%

Type SL – Shrinkage limited cement. The compressive strength at 7 days is 20 MPa and at 28 days 30 MPa and the shrinkage at 28 days a maximum of 750 microstrain when measured in accordance with AS 2350.13.

(Type SL was added to AS 3972 in Amendment No. 1 February 1995)

It can be seen from the above, that apart from the specification of the C_3A content for the Type SR, all of the other specifications are performance based. In the case of Type SR there is a Draft Amendment No. 3 under consideration in which the maximum C_3A content will be replaced by a physical test to measure expansion and a maximum limit of sulphate expansion will be set for Type SR.

Index

Page numbers appearing in **bold** refer to figures and page numbers appearing in *italic* refer to tables.

Abrams' law 62–3
Additives (admixtures)
  for air entrainment 85–6
  blast furnace slag 87, 89–91, 93
  definition 82
  effect on properties of concrete 82–3
  effects on setting of concretes
    accelerators 83–4
    retarders 84
    stabilizers 84–5
  fly ash 87, 88
  to mortar 51–4
  plasticizers 95
  pozzolanas 87–94
  superplasticizers 94–7
Af_m (alumina iron oxide monosulphate) *40*
Af_t (alumina iron oxide trisulphate) *40*, 41, 42, 57
Aggregates
  aggregate cement bond 57, **58**
  aggregate materials
    dense 58
    fire resistance 170, 171–2
    harmful impurities in 61
    lightweight 59
    particle characteristics 60
  in sulphur concrete 198
Air entrainment
  admixtures for 85–6
  effect on
    bleeding 85
    frost durability 85
    segregation 85
    workability 86
  of polymer–modified cements 185, 187
  test for, in concrete 74
Alite, *see* C_3S
Alkali-carbonate reaction 153
Alkali-silica reaction 151–152
Alumina modulus 30, 31
Australia
  cement and concrete manufacture 9–10
  production of microsilica 102

Belite, *see* C_2S
Blast furnace slag
  additions to concrete, effect on
    attack by soft water 148
    hydration heat evolution 93
    hydration rate 93
    strength development 92–3
    workability 92
  aggregates 58–9
  in expansive cements 192
  in high-performance concrete 105
  production 58–9, 89–90
  in supersulphated cement 189
  use in nineteenth century 8
  use in production of cement clinker 23
Bleeding and segregation
  and air entrainment 85
  of concrete 63, 109–11
  and corrosion of reinforcement 110
  effect of microsilica 111
  of polymer-modified cements 186

and porous surface layer 110
test for, in concrete 74
Bogue equations
   calculation of cement phases 29, 70
   use in classification of cements 47

C_3A (tricalcium aluminate)
   hydration of
      in absence of gypsum 39, **41**
      in presence of gypsum 38–9, 40, **41**
   impurities 36
   morphology of hydration products 41–2
   in Portland cement clinker 23
C_4AF (calcium alumino ferrite)
   hydration 41
   impurities 36
   morphology of hydration products 42
   in Portland cement clinker 23
CSH (hydrated calcium silicate gel)
   effect of microsilica on formation 104, 106
   from hydration of C_2S 38
   from hydration of C_3S 36–7
   morphology 41
C_2S (dicalcium silicate, belite)
   hydration 39
   impurities 36
   in Portland cement clinker 23
C_3S (tricalcium silicate, alite)
   decomposition 32
   effect of microsilica on hydration 104
   hydration 36–8
   impurities 36
   initial hydration reactions 37, **38**
   in Portland cement clinker 23
   rate of heat evolution during hydration **37**
Calcium alumino ferrite, see C_4AF
Calcium chloride
   as accelerator in setting of concrete 84
Calcium hydroxide (**CH**)
   from hydration of C_2S 38
   from hydration of C_3S 36

microsilica and formation of 104
morphology 41
Calcium sulphate dihydrate, see Gypsum
Coloured cement and concrete 190–1
Compressive strength
   of concrete
      Abrams' law 62–3
      aggregate/cement bond 57, **58**, 105
      aggregate materials 58–62
      compaction **64**
      microsilica 105–6
      pozzolana additions 92
      superplasticizers 95–6
      water/cement ratio 62–5, 106
      Feret's law 63
      tests for 75
      type of cement **56**
   of mortar
      effects of water/cement ratio on 52
      effects of mix on 53
      measurement of 73
Corrosion
   of steel fibres in cement and concrete 142, 156–7
   of steel reinforcement in concrete
      depassivation by carbonation 154–5
      depassivation by chloride ions 155–6
      moisture and air 156
      prevention 164
      stress corrosion 156
Creep
   of concrete 113–5
      effect of fibres on creep and shrinkage 132, 138
      shrinkage and 114
      tests for 74
   of high-strength (performance) concrete 115
Curing of concrete 159–61

Dental cements 200
Deterioration of concrete by
   alkali–carbonate reaction 153

INDEX

alkali–silica reaction 151–2
corrosion of steel reinforcement 154–7
frost attack 153–4
sea water 150
soft water 146–8
sulphates 48–9, 148–9
sulphuric acid 151
Durability of concrete
  effect of
    compaction 160
    concrete cover over reinforcement 160
    curing 160
    maintenance 160
    microsilica additions 161–2
    mix design 159–60
    pozzolana additions 87, 161
    superplasticizers 97
  frost attack
    effect of air entrainment 85
  of high performance concrete 106, 162
  measurement of
    accelerated testing 163–4
    long-term testing 163
    use of strength 162
Dicalcium silicate, see C_2S

Ettringite
  from hydration
    of C_3A 39, 40, **41**
    of expansive cements 193–4
    of supersulphated cement 189
  from sulphate attack on Portland cement 48, 148–9
Expansive cements
  control of expansion time 192
  hydration of 193–4
  mechanism of expansion 194–5
  production 192
  types of expansive cements 193–4
  uses 192, 195

Fast setting and hardening cements 183–4
Fatigue of concrete 115–6
Feret's law 63

Ferrocement 8, 130
Fibre-reinforcement
  of cement with
    asbestos 133–5
    carbon 137–8, 143–4
    glass 135–7, 143
    natural 140
    polymers 138–40
    steel 130–3
  classification 128–9
  of concrete with
    polymer 140, 142, 143
    steel 140–2, 143, 144
  corrosion of steel fibres 142, 156–7
  early patents 128
  effect on creep and shrinkage 132, 138
  failure mode 129
  ferrocement 8, 130
  fibre–cement interface (bond) 131, 134, 137
  fire resistance of steel fibre-reinforced concrete 173
  in sprayed cement 132–3, 138
  microsilica additions 133, 137, 138
  properties of fibres *141*
  of sulphur concrete 199
  toughening by 129–30, 131
  uses 142–4
Fire damage to concrete
  assessment of fire-damaged concrete 174–5
  effects of
    heat capacity of concrete 172
    presence of microsilica 172–3
    temperature 171–2
    thermal conductivity of concrete 172
  fire damage on steel fibre-reinforced concrete 173
Fly ash
  classification 88
  production 88
  use
    in high-performance concrete 105
    in sprayed mortar 139

INDEX

Frost attack on concrete 153–4
and air entrainment 85, 187

Gunite, see Sprayed mortar

Gypsum
addition to Portland cement clinker 22
cements 2
effect on hydration of C_3A 38–9, 40, **41**
mortars, use in pyramids 1
sulphate attack 48–9, 148–9
in supersulphated cement 189

High alumina cement (HAC)
clinker
composition 179
structure 179
composition ranges *177*
development 9, 177
failures 182
hydration reactions 180–1
kilns for production 177
raw materials 178
strength development **181**
sulphate attack 177
types 177
uses
building material 181–2
refractory castable 183

High performance concrete (HPC)
containing
blast furnace slag 105
fly ash 105
microsilica 105
superplasticizers 105
compressive strength 106
definition 100
durability 106
fire resistance 172
microstructure 104
plastic shrinkage 111
production 105
rate of hydration 104
sedimentation and bleeding 111
structure 106
water/cement ratio 106

Hydration of
barium alumina cement 197

C_3A
in absence of gypsum 39, **41**
in presence of gypsum 38–9, 40, **41**
C_4AF 41
C_2S 39
C_3S 36–8
expansive cement 193–4
high alumina cement 180–1
lime–pozzolana mortars 92
Portland cement-blast furnace slag 93
Portland cement paste 42–6
Portland cement–pozzolana mortars 92
pozzolanas 93
supersulphated cement 189–90

Hydraulic setting cement
eighteenth century 5
natural 7
Roman cement 3, 5
Portland cement 6

Hydrogarnet
from hydration of C_3A 39, **41**

Kilns for the production of
high alumina cement 178
Portland cement 13–2

Lime-based concrete, origins 1

Lime mortars
containing pozzolanas 3, 92
history 2

Lime saturation factor 30, 31

Macrodefect free cement
production 187, **188**
strength 188
uses 188

Microsilica
additions to
fibre-reinforced cement and concrete 133, 137, 138
sprayed mortars (cement) 104, 132–3, 138
effects on
alkali-silica reaction 151–2
bleeding and separation 111

INDEX

fire resistance of concrete 172
hardening of concrete 103–4
microstructure of concrete 104
rate of hydration of cement 104
pozzolanic properties 104
production 102–3
structure 101–2
Monosulphoaluminate hydrate
  from hydration of C_3A 39, 40, **41**
Mortars
  additives 51–4
  'air-mortar' 2
  gypsum
    use in pyramids 1
    composition 2
  lime
    history of 2
  Portland cement based
    compressive strength 52, 53
    pozzolana additions 92
    sand 52
    setting 52
    sprayed 54, 104, 133, 139
    tests 72–4
    typical mixes 53
    water/cement ratio 52
    water retention 52
    workability 51–2
  pozzolanic
    use by Romans 3
  sand-lime
    setting 51

Non-destructive tests 78–80

Pantheon 3–4
Pigments for cements and concretes 190–1
Polymer-modified cements and concretes
  classification 184
  effects of polymer additions
    adhesion 187
    air entrainment 185, 187
    before setting 185–6
    bleeding and segregation 186
    freeze–thaw (frost) resistance 187
    setting behaviour 186

slump 186
water retention 186
workability 185
production 184
properties and uses 186–7
Portland cement
  aggregate/cement bond 57, **58**
  clinker
    cooling rate 31–2
    grinding 22
    impurities in 31–2
    microstructure 33, **34**
    phases 23
    production in rotary kilns 13–20
    proportion of phases in 26, 29
    quaternary phase diagram 27, **28**
    ternary phase diagrams 23–7
  coloured 190–1
  composition
    alumina modulus 30, 31
    Bogue calculation 29
    lime saturation factor 30, 31
    silica modulus 30, 31
  concrete
    compressive strength 56–64
    deterioration and corrosion 146–58
    durability and protection 159–69
    high performance 100–7
    pozzolana additions 92–4
    resistance to fire 170–6
    workability 64–68
  hydration
    of cement paste 42–6
    of components 36–41
    evolution of heat 42, **43**, 87
    heat of hydration 44
    rate of hydration 43
    volume changes 44–5
  mortars
    cement–lime–sand 51–4
    cement–plasticizer–sand 51–4
    strength 52
    workability 51–2
  origins 6
  raw materials 22, 24, 29
  sulphate attack
    by gypsum 48

INDEX

by magnesium sulphate 48
by sodium sulphate 48
tests for 70–81
types of Portland cements
  classifications *46*
  effect on compressive strength
    56
  properties and uses 47–9
  resistance to fire 170
  white cement and concrete 190
Pozzolanas
  additions to mortars 92
  additions to concrete 92–4, 148
  artificial pozzolanas
    blast furnace slag 89–91
    fly ash 88–9
    rice hulls 89
  natural pozzolanas 87–8
  origin of activity 91
  setting reactions 92
  use by Romans 3
Prestressed concrete 121–5
  post-tensioning of 122
  pre-tensioning of 123–4
  tendons 125
Protection of concrete
  against corrosion of reinforcement
    164
  surface protective treatments 165–8

Radiation shields 192
Reinforced concrete
  corrosion of reinforcement 119,
    154–7
  creep 125
  development 7, 9
  drying shrinkage 125
  elastic deformation 125
  fire resistance 170
  high performance concrete 125
  prestressed concrete
    from expansive cements 192, 195
    grouted 123–4
    post-tensioning 122–4
    pre-tensioning 122
    ungrouted 124
  protection of 164
  quality of concrete 125, 159–60

reinforcement/concrete bond
  118–19
reinforcement materials
  fibre-reinforced plastic 120
  fusion-bonded epoxy coated
    steel 119, 164
  galvanized steel 119, 164
  steels 119, 120, 164
  sulphur concrete 199
  tendons
    fibre-reinforced plastic 125
    steel 121, 125
Rice hulls
  pozzolana from 89
Roman cements and concretes
  hydraulic setting 3
  use of pozzolanas 3
Rotary cement kilns
  production of high alumina cement
    177
  production of Portland cement
    cyclones 16, **17, 18**
    development 13
    dry process 17
    fuel 18–20
    heat exchangers 15, 17, **18**
    planetary coolers 15
    precalciner 17, **18**
    refractories 20–1
    relative size **19**
    semi-dry process 16
    temperature 28
    temperature profiles 15, **16**
    wet process 14

Shotcrete, *see* Sprayed mortar
Shrinkage
  chemical shrinkage 113
  drying shrinkage
    definition 112
    and aggregate/cement ratio 112
    and expansive cements 192,
      195
    and superplasticizers 97, 113
    and water/cement ratio 112
    tests for 74
  plastic shrinkage
    effect of bleed water 111

INDEX

in high performance concrete 111
Silica modulus 30, 31
Slump
  effect of superplasticizers 96
  measurement 66–8, 74
  of polymer-modified cement 186
Sorel cements
  production 199
  reaction products 199
Special cements and concretes
  coloured cement and concrete 190–1
  concretes as radiation shields 192
  expansive cements 192–6
  fast-setting and hardening cements 183–4
  high alumina cement 177–83
  non-calcareous cements
  chemically bonded cements 199–200
    strontium and barium cements 196–7
    sulphur concrete 197–9
    polymer-modified cements and concretes 184–88
  supersulphated cement 189–90
  white cement and concrete 190
Sprayed mortar 54
  additives
    microsilica 104
    fibres 133, 139
Standard tests
  cements
    chemical and mineralogical composition 70
    fineness 71
  cement pastes
    plasticity 71
    unsoundness 72
  concrete, fresh
    air entrainment 74
    bleeding 74
    setting times 74
    slump 66–8, 74
  concrete, set
    compressive strength 75
    creep, 74
    drying shrinkage 74
    tensile strength 76–8
  mortars
    compressive strength 73
    consistency 73
    flexural strength 73
    tensile strength 73, **74**
    water retention 73
  non-destructive tests
    cover depth over reinforcement 78
    delaminations in concrete 78–9
    density 79
    elastic modulus 79
    hardness 79–80
Strontium and barium cements
  hydration 197
  uses 196
Sulphate attack
  on Portland cement 48–9, 148–9
  effect of
    calcium chloride 84
    pozzolana additions 94
  sulphate resistant materials
    high alumina cement 177
    supersulphated cement 189
Sulphur concrete
  aggregates 198
  production 197–8
  properties *199*
  reinforcement 199
  uses 197, 198
Superplasticizers
  effect on
    compressive strength 96
    slump 96
    water/cement ratio 95, 101
    workability 95
  use in high-performance concrete 105
Supersulphated cement
  hydration of 189–90
  raw materials 189
  resistance to
    sea water 189
    sulphate attack 189

Thermal cracking 111–12
Tricalcium aluminate, *see* **C_3A**

Tricalcium silicate, *see* **C$_3$S**

Water/cement ratio
  in concrete
    and compressive strength 62–5
    drying shrinkage 112
    and microsilica 105, 106
    and sedimentation and bleeding 109
    and superplasticizers 95
  in high performance concrete 106
  in mortar 52

White cement and concrete 190
Workability
  of concrete 65–8
  effect of
    air entrainment 86
    pozzolana additions 92
    superplasticizers 95
  measurement by
    flow table 67–8
    slump test 66–7
  of mortar 51–2
  of polymer-modified cement 185